Charles-Edouard Brown-Séquard

Experimental and clinical Researches applied to Physiology

and Pathology

Charles-Edouard Brown-Séquard

Experimental and clinical Researches applied to Physiology and Pathology

ISBN/EAN: 9783337140588

Printed in Europe, USA, Canada, Australia, Japan

Cover: Foto ©berggeist007 / pixelio.de

More available books at **www.hansebooks.com**

EXPERIMENTAL AND CLINICAL RESEARCHES

APPLIED TO

PHYSIOLOGY AND PATHOLOGY.

FROM August, 1852, to August, 1853, I published in the *Medical Examiner*, of Philadelphia, a series of thirty-three short papers, which were afterwards connected in one volume, under the title : "*Experimental Researches applied to Physiology and Pathology.*" The following article is the first of a second series of papers, which, with the preceding series which has appeared in Philadelphia, will form a complete summary of all my original researches in various branches of the medical sciences.

I. ARTIFICIAL PRODUCTION OF AN EPILEPTIFORM AFFECTION IN ANIMALS, AND ETIOLOGY AND TREATMENT OF CERTAIN FORMS OF EPILEPSY IN MAN.

Six years ago, I discovered that certain alterations of the spinal cord, upon mammals, produce, after a few weeks, a convulsive affection, resembling epilepsy. (See *Comptes Rendus de la Soc. de Biol.*, t. ii., pp. 105 and 169—1850.) Since that time, I have found many new facts concerning this affection ; and lately, in comparing the results of my experiments with what has been observed in man, in many cases of epilepsy, I have been led to some conclusions, which are, I think, very important, as regards the etiology, the nature and the treatment of epilepsy. Although some of the results of my experiments have already been published (see my *Exper. Researches applied to Physiology and Pathology*, pp. 36 and 80, the *Archives de Médec.*, etc., Fevrier, 1856 ; and the *Moniteur des Hopitaux*, Oct., 1856, p. 954), I will relate them here, as I shall have to make use of them when I expose my views upon the pathology and treatment of epilepsy. I will also give a detailed account of some of the facts I have observed in animals, because these facts throw a great deal of light upon the phenomena of epilepsy in man.

§ I. I have found that the following kinds of injury to the spinal cord are able to produce epilepsy, or at least a disease resembling epilepsy, in animals belonging to different species, but mostly upon guinea-pigs.

1st. A complete transversal section of a lateral half of this organ.

2d. A transversal section of its two posterior columns, of its posterior cornua of gray matter, and of a part of the lateral columns.

3d. A transversal section of either the posterior columns or the lateral, or the anterior alone.

4th. A complete transversal section of the whole organ.

5th. A simple puncture.

Of all these injuries, the first, the second and the fourth seem to have more power to produce epilepsy than the others. The first particularly, *i. e.*, the section of a lateral half of the spinal cord, seems to produce constantly this disease in animals that live longer than three or four weeks after the operation. After a section of either the lateral, the anterior or the posterior columns alone, epilepsy rarely appears, and it seems that in the cases where it has been produced, there has been a deeper incision than usual, and that part of the gray matter has been attained. In other experiments, few in number, the section of the central gray matter (the white being hardly injured) has been followed by this convulsive disease. I have seen it but very rarely after a simple puncture of the cord.

It is particularly after injuries to the part of the spinal cord which extends from the seventh or eighth dorsal vertebra to the third lumbar, that epilepsy appears.

§ II. Usually this affection begins during the third or fourth week after the injury. In some cases I have seen it beginning during the second week, and even one or two days before. At first the fit consists only in a spasm of the muscles of the face and neck, either on one or the two sides, according to the transversal extent of the injury. One eye or both are forcibly shut, the head is drawn towards one of the shoulders, and the mouth opened by the spasm of some of the muscles of the neck. This spasmodic attack quickly disappears.

After a few days the fit is more complete, and all parts of the body, which are not paralyzed, have convulsions. According to the seat of the injury, the parts that have convulsions greatly vary. When the lesion is near the last dorsal vertebræ or the first lumbar, and consisting of a section of a lateral half of the spinal cord, convulsions take place everywhere, except only the posterior limb on the side of the injury. If the lesion consists of the section of the two posterior columns and a part of the lateral columns, and of the gray matter, convulsions take place everywhere without exception, but with much more violence in the anterior parts of the body. When the lesion exists at the level of the last dorsal vertebræ and consists in a transversal section of the two anterior or of the two lateral columns, convulsions are ordinarily limited to the anterior parts of the body; but it is a very interesting fact that they are not always confined to these parts, the two posterior limbs having sometimes very strong tetanic spasms, at the same time that there are clonic convulsions in the anterior limbs. After a transversal section of the central gray matter, or of the whole spinal cord, in the dorsal region, convulsions are limited to either the anterior or the posterior parts of the body.

§ III. Convulsions may come either spontaneously, or after certain excitations. The most interesting fact concerning these fits is that it is possible, and even very easy, to produce them by two modes of irritation. If we take two guinea pigs, one not having been submitted to any injury of the spinal cord, and the other having had this organ injured, we find, in preventing them from breathing for two minutes, that convulsions come in both; but if we allow them to breathe again, the first one recovers almost at once, while the second continues to have violent convulsions for two or three minutes and sometimes more. There is another mode of giving fits to the animals which have had an injury to the spinal cord. Pinching of the skin in certain parts of the face and neck is always followed by a fit. If the injury to the spinal cord consists only in a transversal section of a lateral half, the side of the face and neck which, when irritated, may produce the fit, is on the side of the injury; i. e., if the lesion is on the right side of the cord, it is the right side of the face and neck which are able to cause convulsions, and *vice versa.* If the two sides of the cord have been injured, the two sides of the face and neck have the faculty of producing fits, when they are irritated. No other part of the body but a portion of the face and neck has this faculty. In the face, the parts of the skin animated by the ophthalmic nerve cannot cause the fits; and of the two other branches of the trigeminal nerve, only a few filaments have the property of producing convulsions. Among these filaments, the most powerful, in this respect, seem to be some of those of the suborbitary and of the auriculo-temporalis. A few filaments of the second, and perhaps of the third cervical nerves, have also this property of producing fits. In the face, the following parts may be irritated without inducing a fit : the nostrils, the lips, the ears, and the skin of the forehead and that of the head. In the neck, there is the same negative result when an irritation is brought upon the parts in the neighborhood of the median line, either in front or behind. On the contrary, a fit always follows an irritation of some violence when it is made in any part of a zone limited by the four following lines : one uniting the ear to the eye ; a second from the eye to the middle of the length of the inferior maxillary bone ; a third which unites the inferior extremity of the second line to the angle of the inferior jaw ; and a fourth which forms half a circle, and goes from this angle to the ear, and the convexity of which approaches the shoulder.

§ IV. Can we attribute to the great degree of sensibility of the face and of the neck the property exclusively possessed by these parts to produce fits in animals which have had their spinal cord injured ? In other words, is it in consequence of the pain felt, that there are fits in these circumstances ? This explanation is quite in opposition with the following facts. 1st. When the injury exists only in one of the lateral halves of the cord, the face and neck on the other side have not the power of producing fits, whatever is the degree of the irritation upon them. 2d. In the same case, the posterior limb on the side where the cord is injured, is in a state of hyperæsthesia, and, nevertheless, the most violent irritations upon this limb do not produce fits. 3d. It is sometimes sufficient to touch the face or the neck, or even to blow upon them, to produce the fits. Therefore, unless we admit that there is an extraordinary degree of hyperæsthesia in the parts which possess the faculty of producing the convulsions when they are irritated, we must admit that it is not the pain which causes these convulsions. There does not seem to be more sensibility in these parts than in other parts of the body. When a fit, or rather a series of fits, have taken place, and when, consequently, the power of having them is much diminished, it is easy to ascertain that these parts seem not to be more sensitive than others. The animal does not cry more when they are pinched or galvanized, than when other parts are irritated in the same way.

The production of fits by the irritation of certain parts of the neck and face, seems to belong to reflex actions. It is well known that an irritation of the skin and of the mucous membranes may easily produce certain reflex movements, which very rarely take place after an irritation of the trunks of the sensitive nerves. For instance, coughing is almost a constant result of an irritation of the mucous membrane of the larynx and of the bronchial tubes, while it is very rarely produced by an irritation of the trunk of the par vagum. Something similar exists for the production of convulsive fits when the face is irritated in animals upon which the spinal cord has been injured. If we lay bare the nerves of the face and neck of these animals, we find that even the greatest irritations upon them do not produce a fit. Besides, if we dissect a large piece of the skin of the face, so as to let it be in connection with the nervous centres only by the suborbitory nerve, we find that the irritation of this piece of skin is still able to produce convulsions, while the irritation of the very nerve which connects it with the brain does not produce any. It seems, therefore, that it is in the cutaneous ramifications of certain nerves of the face and neck that resides the faculty of producing convulsions in the animals upon which I have injured the spinal cord. There is, in that case, as I will show hereafter, something resembling what takes place in man in cases where a ligature around a limb is sufficient to prevent a fit of epilepsy.

§ V. What is the nature of the fits that we find in animals upon which the spinal cord has been injured ? I think these fits ought to be considered as epileptic. The following description of these convulsions will show that, if they are not positively epileptic, they are at least epileptiform. When the attack begins, the head is drawn first, and sometimes violently, towards the shoulder, by the contraction of the muscles of the neck, on the side of the irritation ; the mouth is drawn open by the contraction of the muscles of the neck, which are inserted upon the lower jaw, and the muscles of the face and eye (particularly the orbicularis) contract violently. All these contractions usually occur simultaneously. Frequently at the same time, or very nearly so, the animal suddenly cries with a peculiar hoarse voice, as if the passage of air were not free through the vocal chords, spasmodically contracted. Then the animal falls, sometimes on the irritated side, sometimes on the other, and then, all the muscles of the trunk and limbs that are not paralyzed become the seat of convulsions, alternately clonic and tonic. The head is alternately drawn upon one or the other side. All the muscles of the neck, eyes and tongue contract alternately. In the limbs, when the convulsions are clonic, there are alternative contractions in the flexor and the extensor muscles. Respiration takes place irregularly, on account of the convulsions of the respiratory muscles. Almost always there is an expulsion of fæcal matters, and often of urine. Sometimes there is erection of the penis, and even ejaculation of semen.

These are the features which render these fits very much like epilepsy. But they seem to differ from this disease, by the three following characters : 1st. The animals sometimes cry during the fits, when they are irritated, and it seems, therefore, that they have not lost their sensibility. Now as the loss of sensibility is considered a symptom essential to epilepsy, it appears that we ought not to consider as epileptic the convulsions existing in these animals. But, we cannot admit this as a decisive objection, when we remark that frequently they seem to be deprived of sensibility, and that, in man, during true fits of epilepsy, there are sometimes periods where sensibility is not lost. 2d. These animals usually have no foam at the mouth, and this symptom has been considered by many writers as essential to epilepsy ; but there can be no doubt that there are cases of epilepsy without any foam. Besides, we may easily understand why there is no foam ordinarily in animals : usually their fits do not last long enough. 3d. The fits in these animals are most frequently a series of fits lasting two or three minutes, and separated one from the other by a period of one or two minutes, during which the animals are able to rise and to stand on their feet. In this respect these animals differ from the majority of epileptic men, who have not a recurrence of fits after so short a period of calm ; but there are cases of rapidly-recurring fits in man, and therefore we cannot deny that the fits of these animals are true epileptic fits, on the ground that they have that peculiar character of rapid recurrence.

The apparent differences between the fits in animals which have had the spinal cord injured, and true epilepsy in man, ought not, therefore, to prevent our considering them as epileptic fits. Not only the convulsions resemble those of true epilepsy, but the fits are not mere accidents, and they come by series of two or three, once a week, once a day, or even ten or twenty times a day, and the disease lasts for years. Besides, we find, after long and violent fits, that these animals are, for a time, in a state of drowsiness, like men after epileptic convulsions. It seems rational to conclude, from this discussion, that if the convulsions of these animals are not truly epileptic, they are at least epileptiform.

§ VI. The facts expressed in the preceding parts of this paper lead to many interesting conclusions. *First*, they give a positive proof that an injury to the spinal cord may be the cause of an epileptiform affection. *Secondly*, they show a wonderful relation between certain parts of the spinal cord and certain branches of some of the nerves of the face and neck. *Thirdly*, they show that epileptiform convulsions may be the constant consequence of slight irritations upon certain nerves. *Fourthly*, they show that even when an epileptiform affection has its primitive cause in the nervous centres, some cutaneous ramifications of nerves, not directly connected with the injured parts of these centres, have a power of producing convulsions, which other nerves, even directly connected with them, have not. *Fifthly*, they show that the cutaneous ramifications of certain nerves may have the power of producing convulsions, while the trunks of these nerves have not this power.

§ VII. The constant appearance of a disease very much resembling epilepsy, after certain injuries to the spinal cord, in animals, will perhaps settle the undecided question whether epilepsy, in man, may originate from an alteration of the spinal cord or not. It seems very strange that physicians have been so unwilling to admit that the spinal cord could be the seat of the primitive cause of epilepsy, when they admit that any nerve or any part of the encephalon, being altered, may produce epilepsy. The seat of this disease seems to be together in the part of the brain where resides the faculties of Perception and of Volition, and in the part of the cerebro-spinal axis endowed with the reflex faculty ; but, whatever may be thought on this subject, it seems quite certain, from facts observed in man and in animals, that epilepsy may be produced by various kinds of alterations of the encephalon, of the spinal cord and of a great many nerves. In other words, the peculiar disturbance of the cerebro-spinal axis which constitutes epilepsy, may be generated by alterations of various parts of this nervous axis and by many nerves. This view does not agree with that of the most distinguished among the recent writers upon epilepsy. They have hardly spoken of the influence of the alterations of the spinal cord upon the production of epilepsy. For instance, M. Delasiauve - (*Traité de l'Epilepsie*, 1854, pp. 174–181) does not speak at all of this influence, and we find that he places a case of epilepsy with an hypertrophy of the spinal cord among many other cases forming a series of doubtful or equivocal alterations. Hasse does not pay more attention than Delasiauve to the share of the spinal cord in the causation of epilepsy. He seems to take notice only of the influence of the alterations of the encephalon. (*Krankheiten des Nervenapparates*, 1855, pp. 266–67.) Romberg (*Lehrbuch der Nervenkrankheiten des Menschen*, 3d edition, 1855, vol. i. part 2, p. 686) has written only a few lines on the relations between alterations of the spinal cord and epilepsy. He thinks that some of the facts related by Ollivier d'Angers prove the existence of these relations.

M. Bouchet, who had, in a paper with M. Cazauvielh (*Archives de Médec.*, etc., 1825, t. ix.), mentioned some cases of diseases of the spinal cord with epilepsy, has tried to show in a recent paper (*Annales Médico-Psychol.*, 1853) that epilepsy is usually connected with alterations of the hippocampus major (*cornu ammonis*).

If we take notice of this fact that the spinal cord is very rarely examined, we understand that although the number of cases on record, as far as I know, of alterations of this organ in epilepsy, amounts only to about fifty, there is an immense number of cases in which after death from the so-called idiopathic epilepsy, the brain was examined, but not the spinal cord. In these cases, particularly where nothing was found in the brain, able to account for the disease, it should have been of the greatest importance to examine the spinal cord. Such a neglect is a great fault, particularly since the publication made by Esquirol on the result of his autopsies. In the corpses of ten epileptics, Esquirol (*Traité des Maladies Mentales*, 1838, vol. i., p. 311) found, nine times, various alterations of the spinal cord or of its membranes. In four cases, the spinal cord was softened, particularly in the lumbar region; nine times there were lenticular concretions in the arachnoid, some of which were cartilaginous, some osseous; once there were a great many hydatids in the cavity of the arachnoid.

Mitivie, quoted by Esquirol (*loc. cit.*, p. 311), found concretions in the arachnoid in two children who died from epilepsy.

Two cases of chronic meningitis with epilepsy, have been record-ed by M. Clot. (*Rech. and Observ. sur le Spinitis*, 1820.) One case of this kind is related by Ollivier d'Angers (*Traité des Maladies de la Moelle épinière*, 3ème edit., 1837, vol. ii., p. 319).

Calmeil (*De l'épil. sous le rapport de son siège*, 1824) speaks of four epileptics, in two of whom the spinal arachnoid contained many cartilaginous plates, while in the two others the density of the spinal cord was considerably increased.

Bouchet and Cazauvielh have found, in many cases, circumscrib-ed softenings of the spinal cord, and other alterations of this organ and its sheath.

Forget, quoted by Ollivier d'Angers (*loc. cit.*, vol. ii., p. 571), has seen two very important cases, which have a great analogy with what I have found in animals.

Gendrin, quoted by Ollivier (vol. ii., pp. 502 and 520), has found in two epileptics a tubercle in the cervical region of the spinal cord·

Barthez and Rilliet (*Traité des Maladies des Enfants*, 2d edit., 1854, vol. iii., p. 589) relate a very curious case in which epilepsy existed in a girl, who had an angular curvature of the spine in the dorsal region. The symptoms were very much the same as those existing in my animals, and, as it is in them, there was no foam at the mouth. There was no alteration in the nervous centres, except in the dorsal region of the spinal cord, which was almost liquefied. This softening occupied the whole of the cord transversely, and was about one centimetre long.

I might add many other cases of alteration of the spinal cord in epileptics, recorded by writers of the previous centuries, such as Bouet, Lieutaud, Morgagni, Musel, &c. In the work of Portal (*Observ. sur la Nat. et le Traitement de l'Epil.*, 1827, p. 28) there is a curious case of epilepsy with a dilatation of the central canal of the spinal cord, which was filled with water.

If epilepsy has truly been the result of an alteration of the spinal cord in all or in some of the above cases, it might be asked why there are so many cases of diseases or injuries of the spinal cord without epilepsy. This objection loses its value when we remark that every day there are cases of tumors and various alterations of the brain without epilepsy, and that, nevertheless, no one doubts that this disease is sometimes produced by such lesions. Besides, I have found that certain kinds of injury to the spinal cord, in ani-mals, produce much more frequently than others an epileptiform affection, and there is only one kind of injury which seems to pro-duce it constantly. This injury consists in a section of the whole of a lateral half of the spinal cord. I do not know of a single case, in man, where life has been saved after such an injury had been done to the spinal cord. In some cases, where, probably, a great part of the lateral half of this organ had been divided transversely, there has been no epilepsy. Such a case is recorded by Morgagni (*De sed. & causis morborum*, ep. 53, § 23); another by Boyer (*Traité des Maladies Chirurg.*, 1ère edit., vol. vii., p. 9), and a third by my friend, M. Viguès (*Moniteur des Hôpitaux*, 1855, p. 838). In animals, after an incomplete transversal section of a lateral half of the spinal cord, epilepsy is not very frequently produced. Therefore the negative facts concerning the influence of this injury in man, cannot be considered as a proof that man does not resemble animals in this respect.

I think the following conclusions may be drawn from all that I have said concerning the influence of alterations of the spinal cord upon the production of epilepsy : 1st. There cannot be any doubt that in animals certain injuries to the spinal cord frequently produce an epileptiform affection, if not true epilepsy. 2d. That in man there are a great many cases which seem to prove that alterations of the spinal cord may cause epilepsy.

Now, as we well know that the spinal cord has the same or-ganization and the same vital properties, in animals and in man, it seems, from the first of these conclusions, that it may be stated more positively than I have done in the second, that epilepsy may result from alterations of this nervous centre.

§ VIII. Physicians admit now, two kinds of epilepsy, one of centric and the other of peripheric origin. I will try to show that although it seems to be of peripheric origin, it may, in some cases, be in reality of centric origin.

In animals, after an injury to the spinal cord, if we did not know that this injury exists and is the first cause of the disease, we should be led to admit that it is of peripheric origin, in finding that an irri-tation upon a very limited part of the spine produces fits. In a very important case of epilepsy recorded by Odier, the same thing has existed as in my animals. For many years the disease seemed to be of peripheric origin, and the autopsy has revealed that this was a mistake. This case is so interesting, in many respects, that I will give here a summary of its principal points.

Case I.—A man had frequent *cramps* in the little finger of the right hand. The contractions went on increasing in extent and frequency; they by degrees extended to the fore-arm, the arm and the shoulder, always beginning in the little finger. At last they arrived at the head, and then true fits of epilepsy, with loss of consciousness, took place. By means of two peculiar ligatures round the arm and the forearm, and which the man could tie easily, when he felt the first contractions of the little finger, the attacks were prevented at every threatening for two or three years. Unfortunately, one day he eat and drank too much, and, being intoxicated, he forgot the ligature when the initial cramp appeared, and then he had a violent fit. From this time the ligature had no more influence over the fits; they became very frequent and always began in the little finger. Paralysis came on, and the patient died in coma. *Autopsy.*—An enormous tumor was found in the left side of the brain, below a place where the cranium had been wounded long before (*Odier, Manuel de Médecine Pratique,* 2de edit., 1811, p. 180).

This case and the facts observed in my animals, positively show that the apparent outside origin of epileptic fits does not prove that there is not an organic cause in the nervous centres.

§ IX. There is a great analogy between the aura epileptica, in man, and the pain originating in the skin and face of my animals. In them, as well as in man (when there is a real aura), the trunks of the nerves seem not to possess the faculty of producing fits, whereas their ramifications in the skin, or in the muscles, have this power. In my animals, as well as in man, if there is an interruption of nervous transmission between the skin and the nervous centres, fits are no more seen, or at least their number is very much diminished. I have collected many cases of epilepsy with an evident aura epileptica, in which there has been either a diminution of the fits, or more frequently, a complete cure, after the interruption of nervous transmission between the starting-point of the aura and the nervous centres. In these cases, the following various means have been employed with complete or partial success, either against the aura epileptica or against its production : 1st, ligature of a limb or of a finger ; 2d, sections of one or many nerves, and amputation of a limb, or of other parts of the body ; 3d, elongation of muscles which are the seat of the aura ; 4th, cauterization, by various means, of the part of the skin from which the aura originates.

1st. *Cases of application of a ligature, as a means of preventing epileptic fits.* The cases of cure of epilepsy by the application of a ligature are very numerous. Pelops, a teacher of Galen, seems to have been the first physician who employed a ligature to prevent epileptic fits. Here is a summary of the relation given by Galen of the case of Pelops :

Case II.—An intelligent young man, who did not lose his consciousness during his fits, had a sensation originating in one of the extremities, and ascending from thence to the head. His physician, according to the advice of Pelops, applied a ligature in the middle of the limb, above the part first affected. By this means, the fits did no more take place, although previously they used to come every day. (Galen. *De Locis affectis*, lib. iii., c. 7.)

Faventinus (quoted by Herpin, *Du pronostic et du traitement de l'epilepsie*, p. 393) speaks of an old man who had the aura beginning in a finger. After the application of a ligature round the finger, he was cured. Daniel Pucrari, and Salmuth (quoted by Herpin, loc. cit., p. 398), relate cases in which a ligature round the leg prevented the fits from taking place.

Bonet (*Sepulchretum*, 1700, sect. 12, *De Epilepsia*, appendix, p. 292) relates a very interesting case, of which the following is a summary.

Case III.—A man, 50 years old, at times had a swelling in the groin. From this place, a sensation of prickling slowly descended to the sole of the foot. When arrived there, the sensation rose quickly to the brain, of which it attacked only one side, so that convulsions took place only in the left side of the face and body. The patient did not lose his consciousness, but his speech was altered, because the tongue had convulsions. The patient used to say (but with difficulty), "look, how this atrocious disease torments me." A ligature was applied above or below the knee, as soon as the swelling and the sensation appeared in the groin. He always succeeded, by this means, in preventing the fits, until one evening, when not having been able to place the ligature in good season, he had such a violent fit that he died.

Camerarius (quoted by Herpin, *loco cit.*, p. 403) relates a very important case, which shows that in man, as in my animals, when epilepsy begins, the aura epileptica may at first produce only convulsions of some of the muscles in the neighborhood of the starting point of the pain. The irritation of the skin of the face or of that of the neck, in my animals, as I have said, in §II., excites convulsions only in the muscles of the face and neck on the irritated side. It is a local spasm, by a reflex action, such as takes place often in the muscles of the stump of an amputated limb. The case of Camerarius is very important in this respect.

CASE IV.—A young man, in February and March, 1694, had a spasmodic movement of the left middle finger, and, quickly after, the other fingers and the hand had the same movement. This spasm came on every four, five or six days, without any other trouble in the patient's health. On the 5th of April, while he was showing this spasm to his sisters, and laughing at it, suddenly a convulsion of the whole arm took place, and he fell down in a violent fit of epilepsy. From this time, fits came every three or four days, or after a greater interval. They were always preceded by a spasmodic movement, at first of two fingers, then of the other fingers, and afterwards of the hand and fore-arm. From thence the convulsion slowly extended to the arm, and to the muscles of the neck. Then a rotation was produced in the head by this convulsion ; the patient lost his sight, and afterwards his hearing, and at last the attack took place. During the intervals of the fits, the young man was in good health, except that he had a pain in the hand, as if it had been frozen. Ligatures applied near the elbow, with a large band, sometimes prevented the fit.

I will relate, in a moment, a case recorded by Herpin, in which, as in the preceding case, the convulsions were at first limited to the neighborhood of the origin of the aura epileptica. Boerhaave (quoted by Herpin, *loco cit.*, p. 405) gives the history of a young man who at first had spasms and pain in his feet. During two years, these spasms went upwards into the legs and thighs. At last they attacked the right side of the body and the head, and a complete attack of epilepsy came on. The paroxysm was always retarded when a ligature was placed round the right leg.

In cases recorded by Olaüs Borrichius, by Baster, by Burnet, by Ramazzini (quoted by Morgagni, *De sedibus et causis morborum,* epist. 9, §8), by Van Swieten, by Lafler, by Tissot, by Liboschitz, &c., we find that ligatures on the limbs have been more or less completely successful, when there was a true aura epileptica. More recent observers have also been successful in employing ligatures; among them I will cite Esquirol (*Traité des maladies mentales*, vol. i., p. 404) and Gibert, Sandras and Piégu (quoted by Delasiauve, *Traité de l'Epilepsie*, 1854, p. 427-8).

A case reported by Herpin is worth being reproduced here:

CASE V.—A young girl had something like cramps in the two fingers of the left hand. There was a pain in the back of the hand, and a spasmodic flexion of the fingers. This convulsion took place three or four times a day, for two or three days. They ceased for a few weeks, and then came again; and for three days in succession there were two or three a day. In the evening of the last of these three days, the girl was showing to her brother the contraction of her fingers, when her hand closed, the fore-arm and arm were drawn upwards, so that the hand touched her shoulder. A very painful sensation accompanied the spasm of the arm, and from thence it extended to the remainder of the body, and at last the girl lost her consciousness. Either by means of a ligature, or of compression made by her parents' hands, except once or twice, all the threatenings of fits aborted. A treatment of oxide of zinc cured the patient. (*Pronostic et traitement de l'Epilepsie*, p. 71.)

Instead of a ligature, some physicians have employed with success a tourniquet. Cullen (*Elements of Medicine*, §1318) and Tissot have related cases of this kind.

In some of the patients in whom a ligature has been applied with success, this means has at other times failed to prevent the fit. What are we to conclude from this failure? Is it that the ligature was not well applied? or that it was applied too late? or, at last, that even applied very early and very tightly, there are cases where a ligature cannot prevent the transmission to the encephalon of the nervous irritation constituting the aura? Of these three things, the first two certainly may have existed; and as to the third, it seems possible that a ligature, however well applied, will not always interrupt completely the nervous transmission. A fact which I will relate as the first case concerning the section of nerves in epileptics, might perhaps be considered as proving this inefficiency of ligatures.

2nd. *Cases of section of a nerve as a means of preventing epileptic fits.*

These cases are extremely interesting, because they establish

---

positively the curative influence of the interruption of nervous transmission between the part where the aura begins and the nervous centres. The results of the section of certain nerves in my animals, seem, as I will show hereafter, to agree perfectly well with what has been observed in man after this operation.

Case VI.—A servant girl had epileptic fits, preceded by a pain at the extremity of the index of the right hand. A ligature around the fore-arm, and other means, failed to produce any amelioration. The branches of the radial nerve going to this finger were divided during a fit, and the patient was completely cured. (Portal. *Observ. sur le nat. et le traitement de l'Epilepsie*, 1827, p. 159–60.)

In this case, the ligature had failed, but perhaps it was not strongly applied. Portal says it was *un peu forte*, whereas it ought to have been very tight.

Cullen (*Elements of Medicine*, $1318) relates a case of cure by the section of a nerve.

Short (*Edinb. Med. Essays and Obs.*, vol. iv., p. 523) gives the history of a case of epilepsy cured by the section of a nerve of the leg, and the extirpation of a small tumor.

In the following curious case, the nerves, although not cut by the knife, were divided.

Case VII.—A young soldier, who had had epilepsy for many years, became much worse after having been bled in the feet. He had three fits every day at regular hours, at 6 and 9, A. M., and 2, P. M. A feeling of cold used to precede the fit, before the bleeding of the right foot. The physician, M. Pontier, thought that some nerve had been half divided, and that the increase in the frequency of the fits depended upon the irritated state of this nerve. Various experiments were made concerning the influence of ligatures. At first, one was applied over the knee-joint, just before the 2 o'clock fit, which came, nevertheless, but lasted five minutes, instead of twenty-five as usual. The next morning, the same result was obtained. Thinking the ligature was not sufficient, M. Pontier made use of the tourniquet, the pad of which was applied upon the saphena nerve. The fit lasted only one minute and a half. The tourniquet having been loosened too soon, a universal spasmodic trembling took place, which disappeared at once when the instrument was re-tightened. To make sure that the trembling came from the absence of compression, the instrument was loosened again, and the

3

trembling re-appeared, and ceased again when the pressure was
again increased.   Analogous experiments were tried at the time of
another fit, and the result was the same.   When the place where
the bleeding had been made was irritated, the trembling took place
if the pressure was not considerable ; but it ceased when the pressure was increased.   Before and during two other fits, the same facts
were again observed.   M. Pontier then tried the application of a
ligature, not only on the right leg, but also on the left ; the fit then
did not take place.   From time to time there were slight tremblings,
but the patient did not lose his consciousness.   For three days the
same thing was done, and with the same success.   Then M. Pontier decided to divide the two saphena nerves, but thinking that the
bistoury might frighten the patient, and also being desirous of destroying the cicatrices of the lancet wounds, he applied caustic pot-
ash on each of them.   Since this time, the patient has had no fits.
(*Recueil périod. de la Soc. de Méd. de Paris*, No. 79, p. 201, and
Journal Gén. de Méd., vol. xvi., p. 261.)
   This case is very interesting in many respects.   In the first place,
it illustrates admirably the power of the ligatures upon fits.   In the
second place, it shows the influence of ligatures, and of the section
of nerves, in cases where there is no positive aura epileptica.   I
will speak of this point afterwards.
   The cases of amputation for epilepsy which have been followed
by the cure of the patients, are to be compared to those of section
of nerves.   It is, very likely, by the section of the organs of nervous transmission, that an amputation succeeds in curing epilepsy.
Therefore we have thought proper to collect some of the cases of
this kind which are on record.
   The amputation of the great toe has cured a patient whose history is given by Tissot, quoted by Delasiauve.   (*Loco cit.*, p. 430.)
This case is probably the same which Esquirol (*loco cit.*, vol. i., p.
304) relates, without any mention of the physician who treated the
patient.   He says, " A lady having vainly tried many remedies,
was cured by the amputation of the first phalanx of the great toe,
which was the source of an aura epileptica."
   Dr. W. H. Edwards, of Virginia, has recently published a very
interesting case of a cure of epilepsy by an amputation.   Here is
a summary of the case.

   CASE VIII.—For seven years a girl had epileptic fits, at irregular
intervals, sometimes five or six per week, then none, perhaps, for

two or three weeks. In one of her fits she fell on the hearth, and burnt one of her feet. The injury produced by the burn was not cured after five years, during which the fits were very violent. The leg was then amputated a few inches below the knee-joint. This was done in March, 1852, and since then (up to February, 1855) she has never had a return of the fits. (*The Virginia Medical and Surgical Journal*, March, 1855, p. 204.)

There is an interesting case of cure by amputation in the *Medical Examiner*, of Philadelphia (1841, vol. iv., p. 477).

Perhaps we ought to add here the cases of convulsions which have been cured either by the section of a nerve in amputated patients, or by a second amputation; but almost all the cases of this kind are not true cases of epilepsy; they are either more or less local convulsions, or hysteriform convulsions. Besides, epilepsy produced by alterations of the nerves of the stump in amputated patients, is not frequently curable by the section of a nerve, on account of the inflammation of the whole or a great part of the length of the nerve. A case recorded by Mr. Hancock, and another by Mr. Langstaff, are very interesting specimens of this kind. (See the *Phila. Med. Examiner*, July, 1852, p. 468.)

There is a very remarkable case of convulsions treated with success by the section of a nerve, published by Dr. Harris, of Philadelphia. (*Med. Examiner*, 1833, vol. i., p. 2.)

I could relate many cases of tetanus in which the section of a nerve has proved successful, but although these cases bear out the fact I wish to establish, *i. e.*, the influence of the interruption of nervous transmission in preventing convulsions, I will not give them, because I must here confine myself to the subject of true epileptic convulsions.

Besides amputations of limbs, that of the testicles seems to have been successful in curing epilepsy. Joseph Frank relates a curious case of this kind, in which the aura epileptica began in one of the testicles, which was the seat of an ulcer. It was taken off, and eleven years after the operation no more fits had come. (*Praxeas medicæ universæ præcepta*, vol. i., sect. iii., p. 476.)

3dly. *Elongation of muscles which are the seat of the aura.*

The aura epileptica begins in the ramifications of nerves, either in the skin or in the muscles. We have shown that after an interruption of nervous transmission by compression of a section of nerves, the fits are prevented, or are less violent : we will now relate cases showing that when the aura has its seat in muscles, a good means of interrupting the nervous transmission from them, or rather of producing a change in the state of the sensitive nerves of the muscles, is to elongate them.

Gilibert, quoted by Herpin (*loco cit.*, p. 404), relates a case of epilepsy in which the fit began by a pain in the foot ; on the people about the patient taking hold of her foot, and *drawing upon it*, the fit did not come.

Maisonneuve (*Rech. sur l'Epilepsie*, p. 19) gives the history of a girl who had cramps, either in the leg or in the arm. If somebody seized the cramped limb, and extended it with force, the general convulsions did not come.

The same author relates (p. 189) the case of a young man whose fits were prevented by the elongation of the muscles of the arm, which had cramps.

Another remarkable case is recorded by the same writer (p. 195). A man had cramps in one of his legs, and one of his arms, before the fit. If people came quickly enough to draw strongly his arm and leg, the fit did not occur.

Bending the body, or the head, backwards or forwards, to elongate the muscles which have a cramp, has been employed with success by some epileptics ; so it was by a patient spoken of by one of the annotators of Galen (see Herpin, *loco cit.*, p. 396), who used to bend his body forwards. So it was, again, by a patient who, according to Esquirol (*loco cit.*, vol. i., p. 303), used to bend his head backwards, and by this means avoided the fit.

I could give many other facts to prove the influence of elongation of the cramped muscles to prevents fits of epilepsy ; but the preceding are certainly sufficient.

All the means hitherto spoken of, and by which fits are avoided (ligature, section of a nerve, amputation, elongation of cramped muscles), are alike in one thing; they cut off the communication between the aura epileptica and the encephalon. The means I will now speak of attain the same end, but in another way, which is, the destruction of the aura. Both in my animals, and in man, these last means are able to succeed.

4th. *Cases of cure of Epilepsy by Cauterization and other local means of modification of the parts from which originates the aura epileptica.*

There are a great many cases of this kind. They bear out the same conclusion as the cases of section of a nerve, in showing that the fits were caused by a peculiar influence originating from some part of the skin. Cauterization of the skin of the face and neck by the red hot iron, in my animals, seems to cure them, as I will show hereafter. It appears, therefore, that there is something of the same kind in the condition of the skin of the neck and face in these animals, and in the parts of the skin which are the seat of a true aura epileptica in man.

The most varied modes of cauterization have been employed with success against the aura epileptica. Blisters, moxas, potential cauteries, issues, Dippel's oil, a decoction of ruta graveolus, and various other rubefacients, have been successful in cases reported by Locher, Baster, Dovinctus, Brunner, Stuerlin, Henricus ab Heer, Benzi, Portal, Récamier, &c.

It is useless to mention any of these cases particularly, because there are so many on record that every one knows some of them.

The application of a moxa or of the red-hot iron is, I believe, the best means of cauterization. At least it is so for animals, and the many cases in which epileptics have been cured by a burn (see Portal, *loco cit.*, pp. 160 and 172) agree in showing the power that burning of the skin possesses. In a case by Tulpius (see Herpin, *loco cit.*, p. 399), the aura came from the big toe, and the patient was cured by deep burnings of this toe with the red-hot iron.

Any kind of change in the skin may be the cause of the appearance of epilepsy or of its disappearance. A man, says Esquirol (*loco cit.*, p. 304), had an ulcer on one of his legs ; epilepsy came on after the cicatrization of the ulcer, and each fit was preceded by the sensation of a cold wind in the cicatrix ; a ligature above the knee-joint stopped the fit. A young man, whose case is recorded by Pouteau (quoted by Portal, *loco cit.*, p. 375), had received a blow on the head, and the wound was cicatrized only a year after ; he was then attacked with epilepsy, and the fits gradually became more and more frequent. After having been a year in this condition, he consulted Pouteau, who opened the cicatrix by the application of the cautery. From this day the fits disappeared ; but the patient allowed the wound to be healed again, and epilepsy returned. It disappeared again, after another application of the caustic.

Perhaps various operations which have been followed by the cure of epilepsy are to be explained in the same way as the many cases related in this paragraph. This is true, perhaps, for a case mentioned by Delasiauve (*Traité de l'Epilepsie*, p. 430), and in which, after the extirpation of an encephaloid tumor in the angle of the jaw, an epileptic patient was cured. This explanation is probably good, also, for some of the cases in which trepanning of the cranium has been successful in epileptic patients. Among the cases of this kind that I know, I take four, almost at random, to show the fitness of this explanation. In one of them, a circumscribed and permanent pain in the head led Dr. James Guild to apply the trephine. The patient was cured. (Delasiauve, *loco cit.*, p. 422.) In another case, Dr. Campbell (*Annales Méd.-Psychol.*, vol. xiii., p. 613) applied the trephine on the cranium of a man who had received a blow, and who suffered a great deal from the wound it had produced. No more fits took place, and four years after the operation the man was still well. In a third case, recorded by Benjamin Travers (*A further Inquiry concerning Constitutional Irritation and the Pathology of the Nervous System*, p. 285), the trephine was applied in a place where the cranium was depressed and painful to the touch. The patient was cured. The fourth case I will give in full, as it has not yet been published, and also on account of its importance. I owe the history of this case to Professor Van Buren, of New York, and I give it just as it has been furnished to me by this distinguished surgeon.

CASE IX.—" A healthy married woman, 26 years of age, received a blow upon the side of her head from the clenched fist of her husband, who was intoxicated. The seat of the injury remained permanently tender to the touch, and about five months afterwards she had an epileptic fit, for the first time. The fits recurred from this time in gradually diminishing intervals, and when she was admitted into the New York Hospital, in March, 1856, about three years after the injury, they occurred almost every day.

" Over the centre of the parietal bone of the right side, a portion of the scalp, about the size of a half dollar, was very sensitive on pressure, but no appreciable lesion could be discovered, except, perhaps, a slight puffiness of the integuments at this point. She suffered much from headache, the pain always commencing here, and seeming to radiate from this tender surface to the rest of the head. Before a seizure of epilepsy this local pain, which was always present, invariably became more intense.

" After watching the patient for some weeks, during which time the fits were evidently becoming more frequent, it was observed that she was worse at her catamenial period. In fact, upon the 5th and 6th of April she had no less than twenty-seven distinct sei-

zures. Her memory and other intellectual faculties were observed
to be decidedly impaired. In other respects her health was good.
Valerianate of zinc was tried in doses of two and three grains three
times a day during a fortnight, but without benefit.

" It was then decided, in consultation, to explore the condition of
the sealp and cranial bone at the seat of pain, and to remove a portion of the bone, if it showed any evidences of disease. This was
done on the 10th of May. The patient was etherized, and a free
crucial incision made through the scalp. The periosteum was found
more than naturally adherent to the bone, the surface of which was
somewhat elevated and roughened over a space an inch and a half
in diameter. This altered portion of bone was removed by two
applications of the trephine ; its inner surface was found to be perfectly normal, but its diploe was obliterated.

" The wound was closed accurately, except at the point where
the incisions crossed, and cold water dressings applied. No fit occurred until the 18th of May, when she had three during the day
and evening, followed by active febrile symptoms, with nausea, and
on the following day an erysipelatous blush appeared upon the forehead. On the 19th and 20th she had three fits, but they were not
very severe. The attack of erysipelas lasted the usual time, and
proved to be rather a severe one. The wound of the sealp healed
kindly and uninterruptedly, and, at the end of the erysipelas, was
entirely cicatrized (May 27th). After the seizure which occurred
on the 20th, there was no return of the epilepsy. The patient
was retained in the Hospital until after a menstrual period, and as
this did not take place at the usual time, appropriate remedies were
employed, but it was not until the sixth week that the catamenia
returned, so that the patient was not discharged from the Hospital
finally until July 10th, having had no fit meanwhile.

" The epileptic fits which occurred on the 18th, 19th and 20th of
May, coincidently with the invasion of the erysipelas, seem to
have taken the place of the usual chill, as her attack commenced
without one ; and they were the only fits which occurred after the
operation of May 10th.

" I have seen the patient twice since her discharge from the Hospital, once within the past month (November), and she is in perfect health, having had no threatening whatever of an epileptic fit
since those which ushered in the attack of erysipelas."

The extirpation of two pieces of altered bone in this ease has
certainly not been the cause of the cure of the patient, as there have

been fits after their removal. We are led, therefore, to admit that the cure was the consequence either of the influence of the erysipelas or of a change that took place in the skin while the wound was healing. There are cases on record where either erysipelas, or some other febrile disease, seems to have cured epilepsy; but this is so very rare, that it is much more probable that in the patient of Dr. Van Buren the cure has been effected by the change that the operation has produced in the skin, just where the blow which had caused the epilepsy had been received. The frequency of cures of this convulsive disease by anything that may produce a change in a part of the skin, which, being injured or the seat of a pain, has caused epilepsy, renders it very probable that in this case the cure has been obtained by the change produced by the operation.

While I think that Dr. Van Buren deserves great eulogy for this bold and successful operation, I nevertheless ought to say that with the knowledge that I have now, that epilepsy originates very frequently in the skin, it would be necessary in the future, in cases like those I have just recorded, to employ various means of cauterization, and particularly the application of a red-hot iron, upon the injured skin, before making use of the trephine. Very likely cauterization, in a number of cases, will prove sufficient to cure.

Perhaps we are authorized to place the cases we will speak of now, among those in which the skin was the source of an aura epileptica.

J. Carron (*Journal Général de Médecine*, vol. xiii., p. 242) relates the following case.

CASE X.—A child, 11 years old, had fits of epilepsy two or three times a week, since he was 2 years old. A feeling of cold, coming from one of the upper extremities, preceded the fits. A ligature having been applied around the arm and tightened at each threatening, the fits were avoided. A small tumor was then found on the first phalanx of the thumb, and to ascertain if this tumor was the cause of the fits, although it did not produce pain, the ligature was placed successively on the hand and on the thumb, and the fits were prevented. An incision was then made upon the tumor, and four very small bodies of hard sebaceous matter were taken out. The wound was excited to give much pus, and healed after thirty days. The child was completely cured, and has never had a fit since.

Portal (*Anatomie Médicale*, vol. iv., p. 247) gives the case of a woman whose fits began by a pain in the thumb. Leduc, a pupil of Portal, extirpated a hard portion of the skin (a bunion, very likely— *un durillon*), and the patient was cured.

A foreign body in the ear had caused epilepsy. Fabricius Hildanus extirpated it, and the patient was cured. (Esquirol, *loco cit.*, vol. i., p. 303.)

Esquirol says (*loco cit.*, vol. i., p. 303) : " Donat attended a nun who felt, in the beginning of the fits, a pain in the right mamma, from which the aura ascended to the brain; if an ulceration took place in the mamma, the fit was prevented."

Although the skin is more apt to produce epilepsy than the trunks of nerves, there are many cases where an injury to the trunk of a nerve has caused this disease. Such cases have been recorded by De Haën, Henning, Larrcy, Romberg (*Nervenkrankheilen*, 3d ed., vol. i., part 2, p. 689) and others. I will relate some cases of this kind to show that for them, as for those in which the aura epileptica originates in the skin, the same principle is true, that an interruption between the injured part and the brain is able to cure epilepsy.

Portal (*Observ. sur l'Epilepsie*, p. 210) gives the case of a man who had had a nerve injured in the arm. Convulsions, with loss of consciousness, came on many times. A greater incision was made where the wound existed, and the patient was cured.

The same writer (*loco cit.*, p. 156) speaks of a man who had received a pistol shot in the neck, and who had become epileptic. After some time an abscess was formed in the neck ; one of the shot came out, and the patient was cured.

Dieffenbach (*Die Operative Chirurgie*, vol. i., p. 852) relates the case of a young girl, whose hand had been wounded by a piece of bottle glass. Neuralgic pains, epileptic fits and contraction of the limb had been the results of the wound. The cicatrix was opened, and a small bit of glass was found near a nerve which had been divided by it, and which was swollen and hardened. After the operation the neuralgia, the epilepsy and the contraction vanished, and the girl was completely cured.

Fizes, according to Portal (*loco cit.*, p. 157), has seen a man who had become epileptic after having been wounded by a sword near the great angle of the eye, and who was cured after the extirpation of a small part of the point of the sword which had staid in the wound.

4

Cases more or less resembling the preceding have been reported by Lamotte, Van Swieten, Sauvages, De Haën, Burserius, Lamorier, &c.

Darwin reports that he once saw a child who frequently fell down in convulsions. A wart was found on the ankle, which was cut off, and the fits never recurred.

Epilepsy caused by the irritation of the dental nerves, and cured by the extirpation of some teeth, or by the lancing of the gums, is not uncommon. Some interesting cases of this kind have been reported by Portal (*loco cit.*, p. 205 and elsewhere).

I shall not speak here of the cases of epilepsy produced by an irritation of a mucous membrane, or of a viscus, and which have been cured by the removal of the irritation. These cases are very numerous, and they also prove that epilepsy may be cured by the suppression of the irritation of nerves, either in their peripheric ramifications or in their trunks.

§ X. In the preceding parts of this paper I have given a summary of two series of facts: experiments upon animals, and pathological cases observed in man. I have now to compare these two series of facts one to the other, and to draw conclusions from the results of this comparison.

There is one thing which seems to be quite proved by this comparison: it is that the convulsive affection produced by certain injuries to the spinal cord is true epilepsy, or at least an epileptoid affection. I have shown already that the symptoms (see §§ II. and V.) lead to this interpretation. But this is not all; the greatest analogy exists between what we know of the aura epileptica in man (see § IX.) and what I have found concerning the property that the skin of the face possesses of producing fits in my animals (see § IV.). In them it seems that the face is the starting point of a true aura epileptica, and that, as well as in man, an interruption of nervous transmission between the starting point of the aura and the cerebro-spinal axis, seems to cure epilepsy. The same result seems also to be frequently obtained by either burning or other means of cauterization of the skin in the part from which originates the aura. In these animals, as in man, in the cases we have related, the convulsions seem to take place by a reflex action. In these animals also, as well as in man (for instance, in the case of Odier, § VIII.), although the primitive cause

of the affection is in the nervous centres, there is an aura epileptica coming from the skin, and the interruption of nervous transmission from the skin to the cerebro-spinal centres seems to have
been sufficient, for a time, to prevent epilepsy. Besides, the development of epilepsy in many cases in man is similar to what
takes place in my animals: the convulsions at first are limited to
a few muscles around the starting point of the aura epileptica;
they then extend gradually to many others, and, at last, attack
almost the whole body.

If these analogies prove that the convulsive disease which is
produced in animals by an injury to the spinal cord is epilepsy,
we are led to conclude that in man, also, epilepsy may be caused
by a disease of this nervous centre. This gives a new weight to
the great probability that epilepsy has been the result of alterations of the spinal marrow in at least some of the cases (see §VII.)
where this organ has been found altered in epileptics.

It will perhaps seem strange that we speak only of a great
probability, while some physicians consider the question of the
production of epilepsy by a disease of the spinal cord as quite
decided, and describe a spinal epilepsy as a distinct form of this
affection. I deny the existence of this species of epilepsy, as it
has been characterized by many German writers and by Dr. J.
Copland; and I consider as a fanciful description the pathological
and symptomatic history of this form of epilepsy given by Joseph
Frank, Harless, Schoenlein, Dr. Copland, Canstatt, Colson and
Wunderlich.

Dr. Copland says (*Dict. of Pract. Medicine*, 1844, vol. i., art.
Epilepsy, p. 793) that the spinal epilepsy generally arises from
injuries and concussions of the spine, from caries of the bodies of
the vertebræ or inflammation of the intervertebral substance, and
from inflammation of the membranes of the cord, or effusion
of fluid within the sheath; from the metastasis of rheumatism, or
the disappearance of eruptions, &c. According to Schoenlein and
others, it arises frequently from excess of sexual excitement, and
particularly from onanism. Sometimes it is preceded by great sensibility, formication or irritation of the skin. The fits are generally
characterized by severe convulsions, seminal emissions, and expulsion of urine and fæcal matters. The head is seldom so much
affected as in cerebral epilepsy, and the seizures often approach
nearly or altogether to simple convulsions. One or other of the

limbs is frequently weak, and sensation in them occasionally diminished or otherwise altered during the interval (Copland).
According to the German physicians the convulsions resemble
those of tetanus, and attack mostly the extensor muscles; clonic
convulsions are rare. Besides, there are symptoms of diseased
spine, and particularly pain under pressure in some points.

Dr. Copland believes that disease of the spine, associated with
disease of the uterine function and epilepsy or convulsions, is
not rare. He says, also, that in epilepsy depending upon injury
of nerves, the paroxysm, as in the spinal variety, is rather one of
convulsions than of complete epilepsy (*loco cit.*, p. 793).

The same writers describe as another distinct kind of epilepsy
what they call the cephalic or cerebral epilepsy, in which convulsions are mostly clonic, and not so violent as in the spinal variety,
and the loss of consciousness is the prominent symptom.

In their description the German writers and Dr. Copland have
confounded three distinct things: *first*, cases of disease of the
spine, or its contents, with convulsions (and not epilepsy); *second*,
cases of disease of the spine, or its contents, with epileptic fits,
without loss of consciousness; *third*, cases of disease of the spine,
or its contents, with epileptic fits and loss of consciousness. An
inflammatory disease of the intervertebral substance, or of the
membranes of the cord, &c., is not epilepsy. At first this convulsive affection is not a febrile one, while these inflammations
cause more or less fever; then the fits of epilepsy are separated
by long or short intervals, during which there are no convulsions,
while it is not so in these inflammations, or, at least, the intervals are very short in them; and besides, the disease progresses
quickly towards death or cure. It is wrong, therefore, to call spinal
epilepsy cases of meningitis, &c., in which there are more or less
continuous convulsions and fever.

As to the other kinds of cases, called spinal epilepsy by Copland and others, they do not deserve this qualification, unless we
call them so because epilepsy *seems* in them to be caused by a
disease of the spine or its contents. But there is nothing special
in the symptoms which can lead us to find out that the epileptic
fits depend upon a spinal affection, and not upon a disease either
of the brain or nerves. Of the two kinds of cases: spinal complaint with epileptic fits and conservation of consciousness, and
spinal complaint with epileptic fits and loss of consciousness; this

last kind has certainly nothing to distinguish it from the cerebral epilepsy of Copland and others, and as to the other kind it is impossible, also, to distinguish it from the cerebral form, because consciousness may also not be lost in cases of epilepsy due to a cerebral disease.

The symptoms in my animals, in which the primitive cause of epilepsy is certainly an injury of the spinal cord, and the symptoms in many cases of epilepsy in man, where a disease of the spinal cord or its membranes existed, are entirely like those observed in many cases in which the brain was the only organ altered. Still more, in the same patient there may be the symptoms of the so-called spinal epilepsy in one attack, while in the next we find those of the so-called cerebral epilepsy, and *vice versâ*.

In epilepsy due to a cerebral disease, there are, sometimes, all the symptoms attributed by Dr. Copland and others to their spinal epilepsy : violent tetanic spasms, seminal emission, expulsion of urine and fæcal matters, paralysis of one limb and loss of consciousness. For the existence of paralysis of one limb in epilepsy depending upon cerebral disease, I will refer to a paper of M. Bravais (*Thése sur l'Epilepsie Hémiplégique*, Paris, 1827), and to the work of Dr. R. B. Todd (*Clinical Lectures on Paralysis, Diseases of the Brain, &c.*, 1854. Lect. xiv. *On Epileptic Hemiplegia*).

On another side I could relate a number of cases in which the convulsions were clonic and consciousness lost, and in which epilepsy co-existed with a disease of the spine or its contents. Some interesting cases of this kind are to be found in the works of Herpin (p. 133–38) and Portal (p. 26 and p. 286). A relation of two cases of disease of the membranes of the spinal cord and softening of a part of this organ, with violent epileptic convulsions and loss of consciousness, is given by M. Pageant (*Rech. sur les causes, le siège et le traitement de l'Epil.* Thése. Paris, 1825. *Obs.* v., p. 22, and *Obs.* xii., p. 33). In one of the cases of tubercles in the spinal cord, recorded by Gendrin (see *Traité des Mal. de la Moelle épin.* par Ollivier d'Angers, 3e edit., 1837, vol. ii., p. 502), there were convulsions and loss of consciousness.

It is to be regretted that in a case of alteration of the spinal cord, very much resembling that which most surely produces epilepsy in animals, the symptoms have not been fully described. Prof. E. Geddings, of Charleston, who relates this case, merely says: "Rather a stout man was affected, at frequent intervals, with violent convulsions and much suffering for upwards of eighteen months. In the progress of the case, the convulsions became more violent and recurred at shorter intervals, until he was finally released by death." There was an exostosis of the second cervical vertebra, encroaching so much upon the spinal cord as to produce a complete section of a lateral half of this organ. (*North American Archives of Medical and Surgical Science.* Baltimore. 1835. Vol. I., p. 110.)

In reviewing all the symptoms which exist in epilepsy, not one is found to belong exclusively to epilepsy due to a disease of the brain, of the spinal cord, or of a nerve. Even the existence of the aura epileptica is not a proof that the primitive cause of the disease is in some cutaneous nerves, and not elsewhere. The case related by Odier (see § VIII.) shows that a tumor in the brain, producing epilepsy, may be the cause of an aura beginning in the skin. Another case, recorded by Herpin (*loco cit.*, p. 125), resembles the preceding, as there was an aura epileptica in a girl whose epilepsy was probably due to tubercles in the nervous centres. In my animals there is no doubt in this respect, as the irritation of certain parts of the skin produces fits, although the primitive cause of the epileptoid affection is in the spinal cord. The aura may therefore exist in epilepsy depending upon a disease either of the brain or of the spinal cord, as well as it is known to exist in epilepsy due to alterations of cutaneous or other nerves.

I have had a direct proof that the symptoms of epilepsy depending upon an alteration of a nerve could be exactly the same as those existing in epilepsy due to an alteration of the spinal cord. In a guinea pig in which one of the toes had been bitten, there were fits entirely similar to those which are found in animals of the same species after an injury to the spinal cord, and the fits ceased after a section of the sciatic nerve.

The comparison of what I have seen in animals with what has been observed by others and myself in man, shows that the symptoms of epilepsy cannot indicate whether it originates from a disease of the brain, of the spinal cord, or of a nerve. But it is true, nevertheless, that if together with epilepsy, there are positive symptoms depending upon a disease of either of these organs, it will be very probable that epilepsy itself depends upon this disease. The careful examination of the symptoms which co-exist with epilepsy is, therefore, extremely important, because by them we may find whether this convulsive affection is due to a disease of a nerve, of the spinal cord, or of the brain, and this knowledge is of the greatest value for the prognosis and the treatment.

§ XI. I have been led to believe, by what occurs in animals after an injury to the spinal cord, and by some cases observed in man, that the existence of a particular spot capable of producing fits, when irritated, is not rare in epileptic patients. This spot may or may not be the starting-point of an aura epileptica.

In the interesting thesis of M. Bravais *(Rech. sur les sympt. et le traitement de l'Epilep.*, Paris, 1827, p. 18), there is a case of a man who had fits when he touched himself, or was touched by other persons, on the region of the temporal bone of the right side.

Fernel, according to Esquirol *(Loco cit.*, p. 302), saw epilepsy produced each time pressure was made on the upper part of the head.

Rondelet (*Méthode curative des Maladies*, p. 137) relates the case of a man who had a fit every time his ears were exposed to cold.

In a young man in whom there was an aura epileptica starting from the left hypochondrium, a simple pressure on this region was sufficient to cause the fit (Tulpius, quoted by Portal, *loc. cit.*, p. 180).

While I was lecturing on this subject in Boston, in November last (1856), Prof. E. H. Clarke told me that he had seen a fit of epilepsy produced by pressure upon one of the mammæ.

I have found that irritation of certain parts of the skin by galvanism caused fits in two epileptics. In one of them it was the skin of the bend of the elbow, and in the other the skin of a portion of the neck and face. There was no sensation of an aura epileptica in these two cases.

Probably in many cases, without the feeling of an aura epileptica, and even without a feeling of pain arising from any part of the skin, the fits are caused by a peculiar and unfelt kind of irritation, originating from some part of the skin, or from the sensitive nerve of a muscle. Perhaps it will be possible to detect the existence of such parts of the external tegument, or of such nerves, by various means, of which we will speak hereafter. It is certainly impossible to admit that the sensations which exist when there is an aura epileptica are always the causes of the fits, as we know that sometimes they consist only in a feeling of cold, or a kind of tickling or formication, or a slight pain. Such sensations are certainly unable to produce fits, and therefore there must be some other kind of irritation, not felt, existing together with these sensations, starting from the same point, and producing the fit. Consequently, what is essential in the aura epileptica is not what is felt, but an unknown kind of irritation. This special irritation, we repeat, may exist alone, *i. e.*, without any kind of sensation. It is the essence of an aura, without any feeling. A good illustration of this view may be found in some cases recorded by M. Pontier (see above, § IX., CASE VII.), J. Frank, and Henricus ab Heer. In the curious case we owe to M. Pontier, there was no pain arising from the feet, and nevertheless it is certain that an irritation sprang from them, as we find that the fits were prevented by the application of a ligature round the legs, and afterwards by the section of the saphena nerves. In the case mentioned by J. Frank (*Praxeos medicæ universæ precepta*, vol. i., sec. 3, p. 476), epilepsy had come after a disease of the testicle; the scrotum was much contracted during the fit, and although there was no feeling of an aura, castration was performed, and the patient cured. It is evident that in this case the fits were due to an unfelt aura arising from the testicle. In the case by Henricus ab Heer (cited by Sennert, *Opera Omnia*, vol. ii., p. 489), a young girl had no feeling of an aura epileptica, but as she rubbed her big toes one against the other during the fit, applications of butter of antimony were made upon them, and the patient was cured. It seems that in this case, also, there was, as cause of the fits, an unfelt irritation arising from the toes. It is well known that worms in the bowels may cause epileptic fits, although they sometimes do not give pain or any other sensation. The irritation producing the fits is then unfelt, as in the preceding cases.

On one side, therefore, we find that an irritation coming from the skin or a mucous membrane may produce fits, without being felt; whereas on another side, when there is the feeling of an aura epileptica, the variety of the sensations, and their feebleness, often show that it is not they which cause the fit, so that we must admit that even then it is a peculiar, unfelt irritation which produces the attack. In my animals, as I have tried to prove in § IV., it is no the pain caused by pinching the skin of a part of the face and neck which produces the fit, but a peculiar kind of irritation. Perhaps the special irritation which generates a fit gives sometimes a sensation quite special also, and which cannot be described. Many epileptics speak of a strange and inexplicable sensation. M. Delasiauve thinks he has been enabled to judge upon himself how a sympathetic fit is produced. He had a sore throat, with an engorgement of the cervical ganglions. The least pressure upon these inflamed glands caused a sudden bewilderment (*éblouissement*). The experiment, repeated twenty times, always gave the same result. M. D. says that if the pressure had been continued he would have fainted, and that there was quite a special sensation, progressing as quickly as a flash of lightning from the diseased spot to the head (*loco cit.*, p. 33–34).

In the cases of epilepsy in which there is an unfelt irritation arising from the skin, and producing the fits, is it because the irritation causes immediately a complete loss of consciousness, or because it has not the power of giving sensation, that it is not felt? I cannot answer this question positively. I can only say that it is probable that the two things exist.

If we take notice of these three sets of facts—1st, that there are cases of epilepsy in which an irritation arising from the skin, or from the neighboring parts, may cause fits without being felt; 2dly, that by pressure or galvanization we may produce in a part the kind of unfelt irritation which causes fits; 3dly, that such a part being found, epilepsy may be cured by either the application of ligatures, the section of a nerve, or cauterizations, &c.; it becomes evident that it is of the greatest importance to try to find out, in epileptics who have no aura epileptica, if there is not a part of the skin or of a muscle from which arises an unfelt irritation causing the fits. To ascertain the state of things in this respect, various means may be employed. If the fits are frequent, and if they come at regular times, it will be found, by placing tight ligatures around the limbs, whether the attacks are due to an irritation coming from these parts, or not. Among other means of detecting the existence or absence of a peripheric irritating cause of the fits, I will point out particularly the following: pressure upon the various parts of the body; the application of localized and powerful galvanic currents; the application of ice and of a wet and

5

warm sponge, &c. If any part is the seat of a pain, even if this pain seems to have no relation with the fits, it will be necessary to ascertain whether pressure, galvanism, &c., applied upon this part, produce an attack. If it is in a limb that a pain exists, a ligature will decide the relation of the painful spot with the fits. In cases where there is a cramp in some of the muscles, or in one only, at the beginning of the fit, the inducement of a cramp by galvanism might decide if the attack is due to the irritation of the sensitive nerve of the contracted muscle, or if the cramp is nothing but a manifestation of the attack. If the initial cramp exists in a limb, an elongation of the contracted muscle, or a ligature, might lead to the solution of the question.*

The danger of producing a fit by the employment of some of the means that I have indicated as good to decide if there is an unfelt irritation arising from the skin, or from some muscle, and causing the fits, is not a reason to prevent our making use of these means, because the existence of a fit, particularly when we are prepared for it, is a small evil in comparison with the great benefit that may be derived from such a trial.

In my animals, nothing in the skin of the face and neck (except a slight congestion, which perhaps is the result of the pinching, and other modes of excitation that I employ) indicates that this part has such a power as that which it alone possesses, to cause fits when irritated. It results from this fact, that it would be quite wrong to decide, *a priori*, that an epileptic man, in whom the skin seems to be perfectly healthy, cannot have fits produced by an irritation of some parts of his skin. Even in such a case, therefore, it would be necessary to employ the various means I have indicated, to decide the influence of the skin on the production of the fits.

§ XII. In many of the preceding parts of this paper I have strongly insisted on the influence of the aura epileptica, or of a peculiar kind of irritation of the peripheric nerves, as causes of epileptic fits. I must now show that I was right in this respect.

Herpin, in his important work which I have so often quoted (*loco cit.*, p. 421), tries to prove that the phenomena of the aura epileptica are nothing but the result of a cramp in one or in more

---

* No one will imitate a surgeon, cited by Portal (*loco cit.*, p. 135), who performed an amputation of one of the toes, because the movements of this toe were very violent during the fit!

muscles, and that this cramp is the first convulsion of the attack. The same view had already been proposed by Prichard, who says that the aura generally is " a convulsive tremor commencing in a limb " (A *Treatise on Diseases of the Nervous System*, Part first, 1822, *Note*, p. 88–89). Herpin has gone farther, and tried to prove that this is always the case. He thinks that the aura epileptica, or, in other words, the first cramp, depends upon a change in the nervous centres, and that the seat of the aura varies according to the place where the change begins in these centres. The cause of the attack, according to this theory, is in the cerebro-spinal axis, and the aura is only a manifestation, an effect, of this cause, and, in consequence, cannot be considered as a cause of the fit.

This theory implies that the so-called sympathetic epilepsy does not exist; so that it is a denial of the peripheric origin of epilepsy.

I cannot understand such a denial, because I think there cannot be any doubt as regards the existence of the sympathetic epilepsy, when we take notice of the immense number of cases of this disease in which it has been produced by wounds or blows in various parts of the body, by neuromas, or other tumors, by dentition or decayed teeth, by foreign bodies, by worms, by calculi and other concretions, by diseases of the skin or of the trunks of nerves, &c. I will merely refer to the works of Portal (*loco cit.*, p. 155–185, p. 204–214), Esquirol (*loco cit.*, vol. i., p. 297–305), Delasiauve (*loco cit.*, pp. 217 and 253), and Romberg (*Nervenkrankheiten*, 3d ed., 1855, vol. i., part 2, pp. 689 and 700), where a great many such cases are reported.

When I treat hereafter of the nature and seat of epilepsy, I will try to show that almost always, if not always, there is in this disease an increased degree of the reflex excitability of the cerebro-spinal axis, and that epilepsy seems to consist mostly in this increased excitability. When a wound, or any of the known causes of the sympathetic epilepsy, produces this affection, it does so principally, if not only, by increasing this reflex excitability. I will show also, hereafter, that there are two distinct influences belonging to the various causes of the sympathetic epilepsy: by one, they produce the disease, or rather, the principal element of the disease, *i. e.*, an increase of the reflex excitability; by the other, they produce the fits. I refer, I repeat, to the writers I have just quoted, for facts proving that they may produce the disease, and I will now only try to show, in opposition to the theory of Dr. Her-pin, that they often cause the fits. I will also try to show that many kinds of felt or unfelt irritation of the sensitive nerves of the skin, or of the muscles, have the same power.

A great many facts are opposed to the view that the aura results always from a cramp. In the first place, if this view were true, the sensation of the aura should always be felt where there are mus-cles, and not in those parts, such as the fingers, toes. skin, mammæ, testicles, ears, &c., where there are no muscles, and where, there-fore, there cannot be any cramp. In taking notice only of cases reported by Herpin himself, in his learned historical account of the aura epileptica, we find that in a number of them the aura originated in the following parts : the little finger (two cases, one by Brassavola, the other by Hollier) ; the thumb (one case by Bou-chet and Cazanvielh) ; a finger (one case by Faventinus) ; the big toe (four cases—two by Tulpius, one by Sylvius, and one by Por-tal) ; a cicatrix on the foot (one case by Puérari) ; all quoted by Herpin (*loc. cit.*, pp. 393, 394, 395, 398, 416 and 417). I might have given a much longer list by taking facts from other writers, ancient and modern. There are many other facts in opposition to the view of Herpin, in his own work, some of which have been ob-served by himself. There are cases in which there was a cramp, but, at the same time, a pain in parts where there was no cramp, and it is remarkable that the patients complained of this last pain only. So it was particularly in two cases observed by Herpin himself (Case xi., p. 70 ; and Case xix., p. 134).

If the view of Herpin were true, the sensations of the aura epi-leptica should be always the same, and always those of a cramp. Instead of such a thing, it is well known that these sensations vary extremely, and that they are described as a feeling of tickling, formi-cation, burning, cold, &c. It would be easy to give a long list of cases in which these sensations have existed. Romberg, who ad-mits two kinds of aura, a sensitive and a muscular one, says that the sensitive aura, in some of his patients, consisted of a feeling of formication in the extremities of the fingers and toes, and in others a tickling sensation around the mouth (*loco cit.*, p. 674).

Herpin says (p. 421–422), that he partially believes that epilep-sy has been cured permanently or temporarily, and that the fits have been prevented, by stretching the limbs, frictions, ligatures, section of nerves, cauterizations, extirpation of parts, amputations,

&c. These facts are certainly in direct opposition to his theory, and he feels much embarrassed about them. He tries, nevertheless, to show that there is no contradiction between his doctrine and these facts. His reasoning in this respect can prove only one thing, which is, that almost all the successful modes of treatment above enumerated are very powerful to diminish or prevent a cramp. But Herpin does not show how or why the prevention of a cramp cures epilepsy. Certainly it ought neither to cure the disease, or even to prevent the fit, as, according to the theory, the cause of the fit is in the nervous centres, and the aura, or first cramp or convulsion, is nothing but one of the effects of this cause. Of course, a cause is not destroyed, or rendered unable to act, because one out of many of its effects is annihilated. Herpin, very likely, has been aware of this inefficiency of his theory, as he tries to show—1st, that besides cauterization, in some cases, powerful remedies have been employed; 2d, that he considers as doubtful some of the cases of cure by the extirpation of a tumor; 3d, that some operations have cured, for the same reason that fever and ague, typhoid fever, variola, &c., have.

I am surprised to find this last argument employed by Herpin, as there is nothing similar in the various operations performed for the cure of epilepsy, and these diseases. The alterations in the blood, and the changes in the nutrition of the nervous system which exist in these fevers may cure epilepsy, but in operations consisting in the application of a ligature round a limb, or in the section of a nerve, or in the extirpation of a tumor, there is nothing capable of altering materially the blood, and the nutrition of the nervous system. As to the other arguments of Herpin, they are valuable, but they apply only to a small number of cases.

To show the incorrectness of the view of the Swiss Physician, it might be sufficient, I believe, to remind the reader of the cases of cure of epilepsy that I have given in a preceding section of this paper (see § IX.). They prove peremptorily that the source of fits of epilepsy may be in the peripheric part of sensitive nerves. The fits were certainly due to an external cause of irritation in cases of epilepsy where they have been prevented by the following means:

1st. Application of a ligature around a limb or finger.

2d. Section of a nerve.

3d. Amputation of a limb, a finger, a toe, or the testicle.

4th. Extirpation of a tumor, a foreign body, or a tooth.

5th. Expulsion of worms, of calculi or other concretions.

Some other facts which I have mentioned in the beginning of § XI. show, in the most direct way, the possibility of the production of a fit by an irritation of the periphery of the sensitive nerves. Together with these facts, I might have spoken of a young man, observed by Zimmerman, and who had a fit of epilepsy every time he practised masturbation (Esquirol, *loco cit.*, p. 301). That an external irritation may cause fits is also proved, without any doubt, by the facts I have almost daily observed in animals for many years; facts described in the first part of these papers. It even seems, from what is observed in these animals, and from various circumstances observed in man, that when there is a cramp preceding a fit, the cramp is nothing but the first effect produced by the irritation of a sensitive nerve. Cramps in some of the muscles of the neck and face are sometimes the only effects of the excitation of the skin of the neck and face in my animals, and when a complete fit takes place, it is almost always preceded by the spasmodic contraction of these muscles. So that if we did not know that there had been an irritation of the skin, we might think that the first phenomenon was the cramp of these facial and cervical muscles. In cases of wounds of a nerve, two diseases may follow, epilepsy or tetanus, but in these two cases the first convulsive phenomenon is a cramp in the muscles in the neighborhood of the wound. Many facts of this kind have been collected by Swan *(A Treatise on the Diseases and Injuries of the Nerves,* new edition) and Pflüger *(Die sensorischen Functionen des Rückenmarkes,* 1853). If the wound was not known to exist in a case of epilepsy of this kind, the local cramp would be considered as the first phenomenon of the attack, while it is only a secondary one. Now, if, as I have tried to show in § XI., there may be an unfelt irritation in the periphery of sensitive nerves, causing fits of epilepsy, it is possible that in cases where a cramp in one or a few muscles is the only thing felt by the patient, the cramp is not the first phenomenon, but results from the irritation of sensitive nerves in its neighborhood. In cases where there is both pain in a part without muscles, and cramps in the neighboring muscles, it may be that the cramps are the result of the irritation of sensitive nerves, causing this pain.

We do not deny that the first cramp in epilepsy may be due to some direct or primitive irritation of the nervous centres, but we do not know any cases of this kind; while, on the contrary, we know many cases where there was, in the beginning of the fits, a local cramp, resulting from a secondary irritation of the nervous centres, *i. e.*, produced by a reflex action, due to the excitation of sensitive nerves near the muscles attacked with cramp.

Romberg says that there are two kinds of aura epileptica; a sensitive and a muscular one. The sensitive consists of various sensations, the muscular consists in a cramp *(loco cit.*, p. 674). This distinction is more apparent than real. If there are cases where cramps are not reflex movements, depending upon the irritation of a sensitive nerve, and in which they result from the direct excitation of some parts of the nervous centres, such cases ought to be distinguished from those where a cramp is connected with an aura epileptica. Besides, when this connection exists, *i. e.*, when either a felt or an unfelt irritation of a sensitive nerve causes a local cramp by a reflex action, before it produces the other phenomena of a fit, the cramp is only apparently an aura.

In reality, there is only one kind of aura epileptica, if we leave to this word the meaning which it has had for centuries, *i. e.*, a local sensation preceding a fit. This sensation in some cases exists without any cramp; in other cases it seems to co-exist with a cramp in the neighborhood of its starting-point.

In cases where the irritation of a sensitive nerve causes a fit without being felt, there may exist a local cramp, but the name of aura cannot be given to this cramp, as it is only the first reflex manifestation of the preliminary irritation, the existence of which may be found out by the means mentioned in § XI.

§ XIII. My experiments upon animals, compared with cases of epilepsy observed in man, throw a great deal of light on what we might call the physiology of epilepsy, that is, upon what concerns the etiology, the seat, and what is vaguely called the nature of this disease. It is easy to show that one or the other of the two series of facts we have to compare, if not both, are in opposition with the various doctrines concerning the production and the seat of epilepsy. A short critical examination of these doctrines will prove the correctness of this assertion.

The time has passed away when men of talent were tempted to place the seat of epilepsy in the pituitary body (Joseph Wenzel), in the pineal gland (Greding), or in the spinal cord (Esquirol,

Rcid). The injuries or organic alterations of these parts, as well as of other parts of the nervous system, may be either the cause or an effect of epilepsy, but none of these parts can be considered as the *essential seat* of this affection. The numerous cases of co-existence of epilepsy and of a disease of the pituitary body, related by Joseph Wenzel *(Beobachtungen ueber den Hirnanhang fallsüchtiger Personen.* Edited by Carl Wenzel, Mainz, 1810), have lost their apparent importance since it has been shown by Romberg *(loco cit.,* p. 685) and others (Rokitansky, Engel and Sieveking, in Handfield Jones's and Sieveking's *Manual of Pathological Anatomy,* 1854, p. 267, *Amer. Ed.)* that the pituitary body may be altered although epilepsy does not exist, and that this neurosis may exist without any apparent alteration in this small organ. There is no part of the nervous centres about which the same argument could not be used.

Many writers have asserted that epilepsy must depend upon a disease of the brain (organic or not), on account of the existence of the cerebral symptoms. It is useless to speak of the authors who have been or who still are unacquainted with the phenomena of reflex actions; I will merely refer for their views to the works of Portal *(loco cit.,* p. 143–155) and Delasiauve *(loco cit.,* p. 27–35). But many physicians of talent, knowing very well what relates to the reflex actions, have considered the brain as the essential seat of epilepsy. Thus, this affection is placed among the so-called cerebral convulsions by very able pathologists, such as Romberg *(loco cit.*), Spiess *(Krankhafte Stör. des Nervensystems,* in Wagner's *Handwörterbuch der Physiol.,* vol. iii., 2d part, 1846, p. 188), Russell Reynolds *(The Diagnosis of Diseases of the Brain, Spinal Cord, &c.,* 1855, pp. 143 and 174), and others. According to Dr. John Simon, "the intellectual changes which precede, accompany or follow the progress of the disease, its concurrence with insanity, and its tendency to dementia, further mark the convoluted surface of the hemispheres as the primary seat of the morbid process" *(General Pathology, &c.,* 1852, p. 152, *Amer. Ed.).*

The modern physiologists agree in admitting that the brain proper (the cerebral lobes) cannot give rise to convulsions when it is irritated, in animals. Surgeons have sometimes had an opportunity of ascertaining that, in man also, the brain may be cut without producing convulsions. But these facts merely prove that

usually the brain proper cannot be excited by our means of excitation. They do not prove that it cannot be irritated by other
kinds of irritation. The cerebral lobes, as being the seat of the
will, are certainly connected with muscles and can produce contractions in them, as our voluntary movements constantly prove.
We think convulsions may result from kinds of irritation (as that
of a poison in the blood, for instance) different from those which
we usually employ in our experiments. On another side it may
be that alterations in the nutrition of the brain, and other causes,
produce a change in the vital properties of this organ, or rather
give it something that normally it does not possess, viz., the property of causing convulsions when it is irritated. That such a
change in the vital powers of the cerebral lobes is possible, we
are led to admit, as we know that other parts of the nervous system may acquire vital properties that they have not in their normal state. For instance, some parts of the sympathetic nerve
seem to be deprived of sensibility, but inflammation renders them
very sensitive; the nerves of tendons seem to be without sensibility, but inflammation renders them evidently sensitive, as
has been definitely proved by the well-devised experiments of
Prof. Flourens (*Comptes Rendus des Séances de l'Académie des
Sciences*, 1856, vol. xliii., p. 639). I might give also as a striking
instance of a change in the vital properties of a part of the nervous system, what occurs to the cutaneous ramifications of certain
branches of nerves in the face and neck, after an injury to the spinal cord in animals.

I must now say that, although I admit the possibility of the production of convulsions by an irritation of the cerebral lobes, I do
not think it is proved that these parts of the nervous system have
actually caused convulsions. The facts mentioned by Romberg
(loco cit., p. 625–27) do not furnish such a proof. They may be
explained by admitting that the convulsions depended upon either
an excitation of the sensitive nerves of the meninges, or upon
pressure on the parts of the encephalon which are known to be
excitable, or upon the disturbance of circulation and nutrition in the
excitable parts of the encephalon. The case of a child *(Pathol.
and Pract. Researches on Diseases of the Brain*, by Abercrombie,
4th Edit., 1845, p. 57), upon whose anterior fontanelle pressure
determined convulsions, and the experiments of Portal *(loco cit.*,
p. 149), which gave similar results, cannot prove anything, because

pressure upon any part of the brain through a small opening in the cranium, acts upon the whole of the encephalon.

Until it is proved that the cerebral lobes have directly caused convulsions, we are not entitled to say that the seat of epilepsy is in them. If it is argued that the brain proper must be the seat of this affection because an idea or a remembrance, or a smell or the sight of certain things, may induce a fit, we answer that these causes of convulsions act in producing an emotion, and that emotions have their seat in the pons varolii and the medulla oblongata, and not in the brain. If it is said that the loss of consciousness implies that the cerebral lobes have something to do with epilepsy, we certainly do not deny it; but what is the relation between these lobes and epileptic fits? How can convulsions, *i. e.*, actions the existence of which imply that a great amount of nervous power is employed—how can they be produced by an organ which has lost its principal function? How can this organ be so active in the production of convulsions just at the time it loses its activity as the organ of volition and perception? How can we admit that an organ assumes actions which it is not known to possess, at the same time that it loses its well-known actions?

Those physicians who maintain that the brain is the primary or essential seat of epilepsy, have too much neglected these difficulties and contradictions. Their only argument consists in saying that it *must* be so because the brain is affected; but we might employ a similar argument to say that many other parts of the nervous centres are the seats of epilepsy because they are evidently affected, and as much as the brain. It is interesting to remark that it is just the same argument that Dr. Marshall Hall employs to show that the seat of this affection is in what he calls the *true* spinal cord. Such arguments in the end amount to simply an assertion like the following: it is so, not because it is proved to be so, or because the facts agree in allowing us to admit that it is so, but because we cannot explain it otherwise. An argument of this kind is never a decisive one; but it has no value, and ought not to be employed, when the facts are not explained by the hypothesis considered as the only possible explanation, and still more (as is the case with the supposition that the seat of epilepsy is in the brain proper) when there are facts in opposition to the proposed explanation.

We will try to show hereafter that the loss of consciousness in epilepsy may be explained otherwise than by admitting that the brain is the seat of this affection, and that the loss of consciousness, whether it exists alone or with convulsions, may be due to an action beginning elsewhere than in the brain.

As regards the state of the mind and of the senses after an attack of epilepsy, it is not and cannot be a proof that the seat of this disease is in the brain, as those disturbed states may result from various circumstances existing during a fit.

We must now say a few words of a theory which pretends to solve the difficulty above exposed, concerning the coincidence of the loss of action of the brain and of an increased muscular action. The estimable author of a singular but interesting work (*Epilepsy, and other affections of the nervous system which are marked by tremor, convulsion or spasm, &c.*, 1854), Dr. C. B. Radcliffe, in this book and in his lectures on Epilepsy (*Medical Times and Gazette*, March and April, 1856), grounds an explanation of this difficulty upon the supposed fact that muscular contraction does not depend upon a stimulus by the nervous system, but upon the cessation of all stimulus. Dr. Radcliffe, after Dugès and others, thinks that muscular contraction is a purely physical phenomenon, dependent on ordinary molecular attraction when the muscle is *not* stimulated. If the muscles are at rest, it is because an excitation comes upon them, preventing the molecular attraction from producing contraction. If a voluntary movement takes place, it is because the will has suppressed the nervous action which prevented contraction. In a fit of epilepsy, convulsions take place, together with the loss of consciousness, because the brain and other parts of the nervous centres lose their powers at the same time, and stimuli being withdrawn from the muscles, they are left to the action of molecular attraction, and therefore convulsions are produced. It is useless to discuss a theory like this, which is in opposition to almost all the known and the most positive facts of physiology and pathology. I will merely say that if the theory were true, we should always see convulsions in paralyzed muscles, and also after death at the time when nerves lose their power upon muscles.

According to another theory, which certainly deserves much more attention than the preceding, epilepsy depends upon changes taking place in the circulation of blood in the brain proper, and in the other parts of the encephalon. The germ of this theory may be found in the works of many writers, and particularly in the remarkable book of Prichard (*A Treatise on Diseases of the Nervous System*, part 1st, 1822), but Henle has done so much for it that he may be considered as its originator. In his admirable work (*Handbuch der Rationelle Pathologie*, vol. ii., 1st part, 1855, 2d ed., p. 181–3 and p. 403; and 2d part, 1854, p. 46) he tries to show that there are two kinds of epilepsy, one attended with plethora, the other with anæmia. In both there is as a cause of convulsions, a pressure by accumulated blood in the vessels of the base of the encephalon. We may understand easily the congestion of the brain in plethora, but it is not so as regards anæmia. Henle explains it in this last case, in admitting that when anæmia goes on increasing, the bloodvessels of the upper parts of the encephalon becoming empty, the others necessarily become more filled, on account of the impossibility of the cranio-spinal cavity containing less fluid (an impossibility well established by Kellie, Abercrombie, J. Reid and others). As regards the loss of consciousness, it is attributed to an excess of blood pressing upon the brain proper, in plethoric epilepsy, and to the diminution of blood in this organ, in anæmic epilepsy.

Although we think that many thanks are due to Henle for the efforts he has made to show the relations between the phenomena of epilepsy and the state of the blood-vessels of the various parts of the encephalon, we cannot adopt his theory.

In the first place, if a congestion in the two distinct parts of the encephalon (the brain proper and the basis of .the encephalon) was sufficient to produce epilepsy, this disease would be much more frequent than it is, and we should not see so often hyperæmia of the encephalon without convulsive fits, and nevertheless powerful enough to cause paralysis, delirium or coma. The great work of Prof. Andral (*Clinique Médicale*, 4th ed., vol. v., 1840, p. 217–292) and almost all the treatises on intermittent fever, and particularly those of Bailly and Maillot, afford decisive proofs of the frequency of cases of encephalic congestion without epilepsy. Henle himself has been obliged to say that *individual disposition* is necessary for the production of this convulsive affection.

In the second place, we object to the theory of the learned German physician, because he gives no proof that the mere mechanical action (pressure) due to accumulated blood in the vessels of the basis of the encephalon, is sufficient to produce convulsions.

In the third place, Henle does not give any clear reason why, in the anæmic epilepsy, the bloodvessels of the brain proper contract, while those of the basis do not; and besidés, except concerning lead disease, he does not say what excites contraction in them.

Theories of epilepsy, entirely at variance with the preceding, have been proposed. In the last century, Saillant (*Expér. sur des Animaux pour découvrir le siège et la cause prochaine de l'épilepsie*, in *Hist. de la Soc. Royale de Médecine*, in 1782 and 1783, p. 88–96), without giving a theory of epilepsy, concluded, from some experiments, that it is easier to cause epileptic seizures in producing alterations in the blood than by irritating the nerves or the brain. Had galvanism been known at the time of the researches of Saillant, and had he employed it to irritate the nervous centres, he would have seen much more violent and lasting convulsions than those he observed after having altered the blood by injections of air, &c. His experiments only show that convulsions may be due to altered blood, a fact well known already before his researches.

One of the most eminent medical writers of our times, Dr. R. B. Todd, has recently proposed a theory of epilepsy, which I must discuss at length, on account of the importance it should have if it were true, and of the value that belongs, necessarily, to any opinion held by such an ingenious and experienced physician.

Dr. Todd says, "I hold that the peculiar features of an epileptic seizure are due to the gradual accumulation of a morbid material in the blood, until it reaches such an amount that it operates upon the brain in, as it were, an explosive manner; in other words, the influence of this morbid matter, when in sufficient quantity, excites a highly polarized state of the brain, or of certain parts of it, and these discharge their nervous power upon certain other parts of the cerebro-spinal centre, in such a way as to give rise to the phenomena of the fit."

Dr. Todd then proceeds to say that a very analogous effect is observed when strychnine is given to a cold-blooded animal. This drug may be administered in very minute quantities for some time without producing any sensible effect; but when the quantity has accumulated in the system up to a certain point, then the smallest increase of the dose will immediately give rise to the so-well-known peculiar convulsive phenomena, observed in this kind of poisoning. Dr. Todd adds: "This, then, is the humoral theory of epilepsy. It assumes that the essential derangement of health consists in the generation of a morbid matter, which affects the blood, and it supposes that this morbid matter has a special affinity for the brain or for certain parts of it, as the strychnine, in the case just cited, exercises a special affinity for the spinal cord. The source of this morbid matter is probably in the nervous system, it may be in the brain itself. It may owe its origin to a disturbed nutrition—an imperfect secondary assimilation of that organ—and in its turn it will create additional disturbances in the functions and the nutrition of the brain."

"According to the humoral theory, the variety in the nature and severity of the fits depends on the quantity of the poisonous or morbid material, and on the part of the brain which it chiefly or primarily affects. If it affect primarily the hemispheres, and spend itself, as it were, on them alone, you have only the epileptic vertigo. If it affect primarily the region of the quadrigeminal bodies, or if the affection of the hemispheres extend to that region, then you have the epileptic fit fully developed."—(*Medical Times and Gazette*, Aug. 5, 1854, p. 129.)

This theory is nothing but an ingenious hypothesis which Dr. Todd proposes, without trying to prove it. The only reason he adduces to support his theory is, that in the renal epilepsy there is very likely a poison in the blood, but as regards the other kinds of this convulsive affection, he does not say any thing which may lead to the admittance of his hypothesis. Feeling that he had no proof of the correctness of his views, he says: "To give a more definite character to the humoral theory, we need to discover a morbid matter in the blood, in variable proportions, in every case of epilepsy. This desideratum has, as yet, been only partially obtained." Dr. Todd alludes here to the influence of the accumulation of urea in blood, in the cases of renal epilepsy. Leaving aside, for a moment, this kind of epilepsy, we may say against the humoral theory of the eminent British physician: 1st, That we do not know any fact in favor of it; 2d, That there are a great many facts in opposition to it.

Not only we do not know any fact favorable to this theory, but its author seems to be like ourself, in this respect, as he does not relate a single one. We have never read or heard that a poison produced in the brain, has been found in the blood of epileptics, and we cannot imagine on what ground a fact of this kind is considered as probable by the author of the humoral theory or rather hypothesis.

To establish the humoral hypothesis on a solid basis, it would be necessary to show : 1st, That there is always a poison in the blood of all epileptics ; 2d, That this poison gradually accumulates in the blood until its quantity has become considerable enough to produce the phenomena of the fit ; 3d, That during or after a fit, this quantity diminishes (because if it were not so, the fit would continue or come again and again, after a very short time) ; 4th, That the nature of the poison varies, so that it acts either on the brain proper alone (producing a mere vertigo), or on the other parts of the cerebro-spinal centre alone, or on the whole of this centre at once ; 5th, That this poison has quite a different influence on the brain proper and on the other parts of the cerebro-spinal centre, destroying the actions of the former and increasing excessively the actions of the latter.

Not only none of these points have been made out, but it seems that no attempt has been made in the way of a demonstration in this respect.

That there is a poison in the blood of epileptics is a fact which, nevertheless, is possible, as there are substances in the blood of every man, healthy or epileptic, which by a transformation or by accumulation, may act as poisons, and be the cause of many of the phenomena of an epileptic seizure ; but it is not known whether the quantity or quality of these substances is changed in epileptics, just before the fits.

There are many facts which are in direct opposition with the humoral theory of epilepsy. Certainly it is so for all the cases in which a ligature around a limb or one of its parts, has prevented a fit, and also for the cases in which epilepsy has been cured by the section of a nerve, by an amputation, by the extirpation of a tumor, a tooth or a foreign body, or by the expulsion of calculi, of worms,* &c. If, in all these cases, there was, as the cause of the phenomena of the seizure, a peculiar influence of some poisonous matter on the encephalon, instead of an irritation springing from

---

* A curious case of hystero-epilepsy, due to larvæ in the frontal sinuses, has been recently published by Messrs. Duménil and Legrand Dussaule. These larvæ, which belonged to five different species, were expelled by the nose, and after their expulsion the patient, who had had violent convulsions for many months, was cured. (See the very useful report on the progress of medicine and surgery, entitled *Annuaire des Sciences Médicales*, par le Dr. Lorain, revu par le Dr. Ch. Robin, Paris, 1856, p. 151.)

certain peripheric nerves, the means mentioned would not prevent the fits, and, still less, effect a complete cure of the disease. If we were to admit that it is a poison which causes the phenomena of the seizure in these cases, we should have to admit also that this poison acts on the peripheric parts of some nerves, and not on the encephalon. But there is no more ground for this last hypothesis than for the preceding, because the presence of a poison in the blood is a mere supposition. Besides, if this would be a reality instead of a gratuitous supposition, it would remain to be explained why this poison does not act in some way or other after the section of a nerve, or the extirpation of a tooth, &c.

The humoral theory is in opposition with many other facts, among which are those proving that an emotion or various other moral causes may produce a fit of epilepsy. For cases showing, without any doubt, the influence of these causes and their relative frequency, I will refer to the works of Delasiauve (*Loc. cit.*, p. 219--22) and Moreau, de Tours (*De l'étiol. de l'épil. et des indications*, &c., in *Mém. de l'Acad. Impér. de Médecine*. 1854. Vol. XVIII. p. 1 *et seq.*).

The facts we have related in § XI., to prove that seizures of epilepsy are sometimes produced by a mere pressure upon or by galvanization of a small part of the skin, are also in direct opposition to the humoral theory. How could a pressure upon the skin produce a fit, every time it is made, if the fits were due only to a peculiar influence of a poison on the encephalon?

The following facts resemble, in many respects, those I have mentioned in § XI., and they also are in complete opposition to the humoral theory: they have been collected by Delasiauve (*loco cit.*, p. 137–38), to show the influence of certain circumstances on some epileptics: 1. A patient could not smell hemp, without having a fit.—(*Tissot.*) 2. In another, the same effect was produced by the slightest odor, even that of broth or of a medicine.— (*Schubart.*) 3. A child had a fit every time he saw something red.—(*Buchner, Tissot.*) 4. A child had an epileptic seizure as often as he heard a dog bark.—(*Van Swiéten.*) 5. The idea of phantoms, which had frightened a boy, when quite young, was sufficient to cause the fits.—(*Maisonneuve.*) 6. In a case, the remembrance of a fright was enough to produce the seizure.—(*Van Swiéten.*) 7. Any word of blame, addressed to two patients, gave them a fit.—(*Delasiauve.*)

The cases in which a physical impression has been the cause of the first attack of epilepsy, may be regarded as less valuable against the humoral hypothesis, than the preceding facts in which, at each return of the cause (either moral or physical), a seizure took place. It might be said to diminish their value that the physical impression occurred just at the time when the poison of the blood was beginning to act upon the brain. But in admitting that such a coincidence has sometimes taken place, we certainly cannot imagine that *in all the very numerous cases of epilepsy*, in which the first fit has occurred immediately after a physical impression, such a coincidence has existed. The works of the principal writers on epilepsy, Van Swiéten, Tissot, Maisonneuve, Cooke, Esquirol, Portal, Copland, Herpin, Delasiauve, Moreau (de Tours), &c., contain too many of such facts for our dreaming of the possibility of explaining the production of epileptic fits, immediately after a physical impression, without attributing at least a share in the causation of these fits, to this impression. The *post hoc, ergo propter hoc*, is a sound reasoning when the number of facts is so extremely considerable as it is here.

In my animals, as I· have already said many times, the fits are produced at every time the skin of certain parts of the neck and face is pinched.* As the seizure in these animals takes place when we desire it, we have there a decided proof that, at least in them, fits may be produced otherwise than by the irritation of a poison on the encephalon.

It results from this exposition of facts that, in animals and in man, fits of epilepsy cannot be considered as always due to the influence of a poisonous matter upon the encephalon. We would not say, however, that they are never caused by a poison in the blood. It seems, on the contrary, not only when there is a deficiency in the urinary secretion, but also when the elements of bile are in great quantity in the blood, or when the functions of the supra-renal capsules are suppressed, that epileptiform seizures take place, owing to the irritation that certain substances, contained in the blood, exert upon some parts of the nervous system. When there is not a free menstruation, and perhaps, also, when the secretion of the skin is stopped, it seems probable that a poisonous

* In October, 1855, I had the satisfaction of showing this experiment to Dr. R. B. Todd himself, in presence of many distinguished physicians, among whom were M. W. Bowman, Prof. L. Benle, Dr. R. H. Semple and Dr. R. Druitt.

7

matter remains in the blood, where it accumulates, and that it participates in the causation of epileptic fits.* Besides, it is certain that some poisons, and particularly lead, are able to cause epilepsy. But many questions are still to be solved, concerning the *modus operandi* of poisons which cause convulsions. I have shown elsewhere (*Experimental Researches applied to Physiol. and Pathol.* New York, 1853, p. 57–63 and p. 113) that these poisons have two modes of action, entirely different one from the other. One of these modes, which is by far the most frequent, seems to consist only in an increase of the reflex faculty of the cerebro-spinal centre. The poisons which belong to this category, according to my researches, are the following: strychnine, brucine, cyanhydric acid, cyanide of mercury, morphine, nicotine, picrotoxine, digitaline, sulphide of carbon, oxalic acid, &c. The other mode of action of poisons producing convulsions, consists mostly in a direct irritation of various parts of the nervous system. I do not know of any other poison, acting exclusively in this way, except a substance existing normally in the blood, which accumulates during asphyxia, and which very likely is carbonic acid. The differences between these two modes of action of poisons are striking. In one of these modes there is no irritation, or at least very little, produced upon the nervous system or the contractile tissues, and therefore there is no convulsion directly caused by the poisons belonging to this category.

It will probably surprise many persons to hear that strychnine, cyanhydric acid, brucine, &c., do not directly give convulsions— but this is a fact; these substances do not seem to have any power of excitation either on muscles, on sensitive and motor nerves, or even on the spinal cord. Perhaps some of the poisons, of which a list is to be found above, have a slight power of excitation on the spinal cord, but they certainly do not cause directly the powerful convulsions which are attributed to them. They act almost only in increasing the reflex power of the cerebro-spinal centre, in such a manner that the least excitation, as, for instance, a voluntary or a respiratory movement, or any other kind of irritation of nerves of the skin or of the mucous membranes, causes convulsive

---

* Very judicious remarks on the subject of the influence of poisonous matter contained in blood, in eruptive diseases, in jaundice, in deranged menstruation, in albuminuria, &c., have been made by Prof. Gunning S. Bedford, in his important work, *Clinical Lectures on the Diseases of Women and Children.* Third Ed. 1856. pp. 437, 475, 502 and 525–34.

reflex movements. We might say that they act in giving to the nervous centres the faculty of causing convulsions *when the centres are irritated, but they do not irritate.* (For the proofs of these views, see my work above quoted, p. 57–63.) On the contrary, black blood, or very likely carbonic acid, seems to destroy the reflex power of the cerebro-spinal centre, but while so acting, it *irritates* violently this centre, and, therefore, causes directly powerful convulsions. This last poison differs also from the preceding in being able to irritate directly muscles and motor or sensitive nerves. (See for this and other influences of black blood, or rather of carbonic acid, my work, already quoted, p. 110–13, and p. 117–24. See, also, the thesis of my friend and pupil, Dr. Brandt, entitled *Des phénomènes de contraction observés chez des individus morts du choléra ou de la fiévre jaune,* Paris, 1855, and my paper on red and black blood in the *London Medical Times and Gazette,* Nov. 17, 1855, p. 492–94.)

There are, therefore, some poisons that cause convulsions indirectly, by increasing the reflex power of the cerebro-spinal centre, and not in irritating them, while there are others which cause convulsions directly by an irritation of the cerebro-spinal centre. In which of these two categories are we to place the poisons, contained in blood in cases of epilepsy, where some secretion (the urinary, the biliary, &c.) is suppressed or much diminished? This is quite an undecided question. Many other things are still to be known concerning these poisons; but we do not intend to examine this subject here. We wished merely to say, that even in cases where there is some ground for the humoral theory of epilepsy, proposed by Dr. Todd, we have no proof that the poison acts as this eminent physician supposes. We will add that even in cases of organic disease of the kidney, coincident with epilepsy, we are not entitled to declare positively that it is in consequence of the accumulation of some of the principles of urine in the blood, that the fits are produced, as it might be that they result from an irritation of the renal nerves, as it is the case when there are calculi in the tubuli of the kidneys without a notable diminution of the secretion of these glands. On another side it is very well known, as Prévost and Dumas, Ségalas, Tiedemann and Gmelin, Mitscherlich, Bernard and Barreswil, Stannius, Frerichs and myself have ascertained many times, that after the extirpation of the kidneys, *i. e.,* when the urinary secretion is as much diminished as possible, convulsions are very rarely produced, and never violent. So that in a case of epilepsy with renal disease, either the convulsions have no relation whatever with the renal affection, or if they have a relation, it is either through the agency of the renal nerves, or in consequence of a transformation of some element of the urine in the blood, as these elements seem to be unable to cause convulsions. It is mostly this last argument which has led Frerichs, in his very interesting work on Bright's disease (*Die Bright'sche Nieren-krankheiten*, Leipzig, 1852), to his so-much-debated theory of uræmia.

As a general conclusion of our discussion of the humoral theory of epilepsy, we will say: 1st, that even in the cases where there is probably a poison in the blood, its relations with the production of fits are not known. 2d, that we are not entitled to consider as due to the elements of certain secretions, remaining in the blood, the epileptic fits which may exist when the glands producing these secretions are diseased. 3d, that there are a great many cases of epilepsy in which the cause of the fit is not in the blood.

Normal blood contains substances which may act like poisons, either after a change in their chemical composition or when their quantity is increased. But very few of these deleterious substances cause convulsions, directly or indirectly; most of them kill without producing phenomena resembling those of a real fit of epilepsy. When the cutaneous perspiration is stopped, after the skin has been covered with a layer of varnish, as in the experiments of Fourcault, Magendie, Becquerel and Breschet, the animal dies, without having epileptiform convulsions.* There is also normally in blood a deleterious principle, the accumulation of which, during a fit of epilepsy, must certainly be the cause of greater violence in the convulsions than there would be if the

---

* Concerning this subject, I have made many experiments, the details of which will be found in another paper. I will merely give here some of the principal results: 1. The glands of the skin in the higher animals, and probably in man also, eliminate a poison; 2. These glands are in many respects analogous to the venom-glands of the toad, the salamander, and also the viper and rattlesnake; 3. When these glands (in the higher animals as in the reptiles) are taken away, or rendered unable to act, the poison that they normally eliminate accumulates in the blood, and usually death occurs quickly; 4. It is wrong, therefore, to say that the venom is not a poison for the animal that produces it; 5. If it seems that the rattlesnake, for instance (and the same thing might be said of the toad, the viper, &c.), is not poisoned by its own venom, this depends upon the fact that when introduced into the blood by absorption, the poison is quickly eliminated by the venom-glands; 6. When these glands have been extirpated, the animals are poisoned by their own venom; 7. The sweat of the dog seems to be much more poisonous for a rabbit than for a dog, and *vice versa*.

quantity of this poison did not increase; we mean carbonic acid, or else some other substance which accumulates in the blood at the same time with this acid. In this respect the theory of epilepsy of which we have now to speak—that of Dr. Marshall Hall—has some relation with the humoral theory of Dr. Todd.

According to Dr. Marshall Hall, epilepsy, when it begins to exist, depends upon an increase of the excito-motor power in what he calls the true spinal cord. He thinks that after a great number of fits, the reverse exists; the patient is in a state of exhaustion, due to the loss of the excito-motor power which accompanies each seizure, while the re-production of this power is not adequate to the loss. He acknowledges, however, that although exhausted, the patient is then in a state of extreme susceptibility to new fits. (See one of his latest publications; *Aperçu du Système Spinal*, Paris, 1855, p. 139–140.) Elsewhere, Dr. M. H. says that an epileptic fit is an excessive excitement of the medulla oblongata, the centre of the reflex actions (*loco cit.*, p. 115). He thinks that the causes of inorganic epilepsy act either directly or indirectly upon the nervous centres, so that the convulsions may be direct or reflex (*loco cit.*, p. 108). The true spinal cord having no spontaneous action, and epilepsy depending upon this nervous centre, the result is that this affection consists only in *excited* actions, either direct or reflex. (*Loco cit.*, p. 206.) Dr. M. H. says, " A spasmodic affection of the larynx has obviously much to do in this disease, as well as in the crowing inspiration, or croup-like convulsions of infants; so much, indeed, that I doubt whether convulsion could occur without closure of that organ." (*On the Diseases and Derangements of the Nervous System*, 1841, p. 327.) The eminent physiologist, however, seems to think now that the closure of the larynx, *i. e., laryngismus*, though essential, is not the only cause of the convulsions of epilepsy. To complete the exposition of his views, we must say that he feels much embarrassed concerning the loss of consciousness. He seems inclined to attribute it to the obstacle to the return of venous blood from the brain.

To sum up the views of this distinguished physician, we will say, 1st, That he places the seat of epilepsy in the excitable part of the cerebro-spinal axis, and more in the medulla oblongata than elsewhere; 2d, That he thinks there is an increased reflex power in the beginning of the disease; 3d, That he admits that the convulsions are the results of the asphyxia caused by the closure of the larynx.

We do not think it worth while to discuss the views of Dr. Marshall Hall; a few remarks are sufficient to show that they do not contain an acceptable theory of epilepsy. In the first place, how can this affection at one period of its existence depend upon an increase of the reflex power, and afterward persist, when, according to Dr. Hall, the reflex power is diminished? How can the intense excitement of the medulla oblongata, in which he supposes that epilepsy consists, explain the loss of consciousness which is so frequent in this disease? As there are cases of epileptic loss of consciousness without contraction of the muscles of the neck, the obstacle to the return of blood from the brain cannot be considered as the cause of the cessation of action of the brain. Besides, how can a cause of increased action of the medulla oblongata be a cause of loss of action in the brain?

But although Dr. Hall has not published an acceptable theory of epilepsy, we think he has done much for it in calling attention to the phenomena of laryngismus and trachelismus. We will show hereafter that the state of asphyxia which depends mostly on laryngismus in epilepsy is, in some respects, a more important fact than Dr. Hall himself admitted.

§ XIV. We have tried, in the preceding part of this paper (see § XIII.), to show the deficiencies of the principal theories of epilepsy. We will now state our own views, but before doing so, we wish to declare that we do not pretend to give here a complete theory of epilepsy; we will merely try to elucidate some of the principal questions on this difficult subject.

I have ascertained upon my epileptic animals that the brain is not essential to the production of epileptiform convulsions. After I have taken away the brain proper, in one of these animals, I find that I can produce a fit almost as easily as before the operation, by pinching the skin of the face and neck. The only difference is, that the fit is not so violent, in consequence of the loss of blood. We find that still weaker convulsions may be caused by pinching the face and neck, if, besides the cerebral lobes, we take away the cerebellum, and even the whole of the basis of the encephalon, except the medulla oblongata and the pons Varolii.

From these experiments it results that, in my animals, epilepsy has its seat in either the pons Varolii, the medulla oblongata, or the spinal cord, or in these three parts together. It is very probable that its seat is in the upper part of the spinal cord, in the medulla oblongata, and the pons Varolii, where the roots of the trigeminal and of the first spinal nerves have their origin. According to some experiments made by Eduard Weber and Dr. R. B. Todd, the faculty of producing epileptiform convulsions does not belong to the spinal cord. E. Weber (Art. *Muskelbewegung*, p. 16, in Wagner's *Handwörterbuch der Physiol.*) says, that the application of an electro-magnetic current to the spinal cord of frogs produces tetanic convulsions, while its application to the medulla oblongata causes alternate contractions and relaxations, as in epileptic fits. Dr. R. B. Todd *(London Med. Gazette*, May 11, 1849) states, that while the convulsions excited by the electro-magnetic current passing through the spinal cord and medulla oblongata are tetanic, the muscles being thrown into a state of *fixed* contraction, those which ensue when the current is transmitted through the region of the meso-cephalon and corpora quadrigemina are *epileptic*, being combined movements of *alternate* contraction and relaxation, flexion and extension, affecting the muscles of all the limbs, of the trunk, and of the eyes, which roll about just as in epilepsy. We have performed similar experiments upon rabbits and frogs, which have given almost the same results. In rabbits, when the current was passed through the pons Varolii and the tubercula quadrigemina, there were alternate movements of flexion and extension, resembling those of epilepsy, but much more extensive. When the current passed through the medulla oblongata, there were tetanic movements of the anterior limbs, with epileptiform convulsions of the posterior limbs; sometimes the anterior limbs also had epileptiform convulsions. When the current passed through the spinal cord, a tetanic spasm was produced. We have found that a state strongly resembling a fit of epilepsy exists after a transversal section of the upper part of the medulla oblongata, which state continues to exist as long as the animal lives. We must not, however, conclude from these experiments that the seat of epilepsy is only and always in one or in all of these parts —the tubercula quadrigemina, the pons Varolii and the medulla oblongata. Pressure upon these parts has often taken place in man without causing epileptiform convulsions, or convulsions of any kind. More than ten of the cases of organic diseases of the encephalon, collected by Abercrombie *(Path. and Pract. Researches on the Diseases of the Brain and Spinal Cord,* 4th ed., 1845, p. 433–457), afford sufficient proof of this assertion. The results of the experiments of Weber, of Dr. Todd, and of our own, are certainly interesting, but they cannot lead to the conclusion that the convulsions of epilepsy in man result *constantly* from some affection of the quadrigeminal bodies (as Dr. Todd believes), or of the pons Varolii and medulla oblongata. It must be remembered that the experiments upon animals are made on healthy nervous centres, and that disease changes the vital properties of these centres. Tetanus, or at least, tetanic convulsions, are sometimes due to diseases of the encephalon, and we have shown already (see § X.) that the nature of the convulsions has not any constant relation with the parts of the cerebro-spinal axis (spinal cord or encephalon), primarily diseased in epilepsy. We know that the muscles animated by nerves arising from the encephalon, or by nerves from the spinal cord, very often exhibit the same kind of convulsions in epilepsy, in tetanus, in hydrophobia, in poisoning, &c. Besides, in a great many epileptics, the first convulsions in an attack are tonic (tetanic), and they are succeeded by clonic convulsions. In other epileptics the fits are sometimes entirely tetanic, and more rarely, entirely clonic in the limbs. In certain animals, Dr. Martin-Magron and myself have discovered (see my *Experimental Researches applied to Physiology and Pathology,* New York, 1853, p. 20) that an irritation of the medulla oblongata caused by tearing out the facial nerve causes convulsions which are partly tonic and partly clonic. Other irritations of the medulla oblongata, of the upper part of the spinal cord, of the pons Varolii and its peduncles, of the tubercula quadrigemina, of the auditory nerve, &c., cause also tonic and clonic convulsions (see my work just quoted, p. 18–23, and p. 99). These facts, and many others, compared to the effects of galvanization, show positively that different kinds of irritation produce different effects, and, therefore, we cannot conclude from the fact that epileptiform convulsions are produced by galvanic irritation of the pons Varolii or other parts of the encephalon, that it is an irritation of these nervous centres which causes epilepsy in man.

If we neglect the nature of the convulsions and take notice only of the parts of the body where they first occur, we arrive at the conclusion that the seat of epilepsy is very variable. Usually, however, the first spasmodic contractions occur in the muscles of the larynx, of the neck, of the eyes, of the chest, of the face, and in the blood-vessels of the brain proper, as we will show hereafter; and as these parts are animated by nerves coming from the encephalon and from the upper parts of the spinal cord, it seems that the seat of epilepsy is usually in some of these parts, if not in all. But the seat of this disease may be in other parts of the spinal cord, as seems to be proved by the production of the first spasmodic contractions in one of the limbs, either the inferior or superior. After the first spasms, all the muscles of the body may be attacked with convulsions; so that if we take notice of the loss of the actions of the brain proper, there is ground for thinking that the seat of the disease is both in those parts of the cerebro-spinal axis where reside the faculties of Perception and Volition, and in those endowed with the reflex faculty; but this view is right only in appearance. We have shown already (see § XIII.) that the loss of perception and volition does not prove that epilepsy has its seat in the brain proper; we will try, in a moment, to show the great probability that a contraction of the bloodvessels of the brain proper, due to an irritation of their nerves in the spinal cord and medulla oblongata, causes the loss of the cerebral faculties; and as regards the increase of the reflex faculty, we will show that a partial and a local increase is sufficient for the production of fits.

Are epileptic fits always the result of an excitation of the cerebro-spinal axis? We think that it is so, but we consider it possible, however, that the excitation may arise from chemical and physical changes taking place in the elements of the nervous centres, in consequence of bad nutrition and other causes. In this case it is just the same thing as if an excitation was produced by a tumor, by a poison in the blood, or by a nervous influence arising from some irritated nerve, &c.

As physiology teaches that an irritation of the simple direct motor side of the cerebro-spinal axis cannot cause general convulsions, we are entitled to consider as reflex the convulsive movements which result from direct excitations of the nervous centres, as well as those which result from irritations coming from peripheric nerve-fibres. The so-called *centric* and *eccentric* causes of excitation of epileptic fits, both act on, or through the sensitive or excito-motory side of the cerebro-spinal centres, and consequently both act on the reflex faculty of these centres, so that they both ought to be called reflex excitations.

We think epilepsy depends in a great measure on an increased reflex excitability of certain parts of the cerebro-spinal axis. We shall no longer speak of reflex *faculty* or reflex *property*, because these words do not express what we mean. In all muscular and nervous tissues we find two distinct properties; a property of producing actions, the force of which may vary extremely, and a property of receiving excitations, which we call excitability. One of these two properties may be very strong, while the other is very weak. Take, for instance, the muscles of cold-blooded animals; when the temperature is very low, their excitability is not very considerable, while their force of contraction is very great. When the temperature is high, on the contrary, the least excitation induces them to contract, but their contraction is without force. Again, if we take an atrophied muscle, we find, sometimes, that it may be excited to contract by a galvanic current too weak to excite contractions in a healthy muscle, while if we apply a strong stimulus to both, we find that the healthy muscle contracts with much more force than the atrophied one. Many experiments, which we will publish in another paper, have shown us that the reflex faculty of the cerebro-spinal axis is composed, as the muscular contractility is, of two elementary vital properties, one of which we call the *reflex excitability*, and the other the *reflex force*. The cerebro-spinal axis may have a great reflex force, and very little excitability. It may, on the contrary, have an excessive reflex excitability with very little reflex force. In almost all epileptics, if not in all, the reflex excitability is increased, while the reflex force is rarely above, and often below its normal degree. The reflex excitability may not be much increased, and nevertheless be sufficient for the production of the fit, when certain excitations exist. I have found in my animals that there is not a great increase of the reflex excitability of the cerebro-spinal axis, except in a part of the spinal cord which is separated from the rest, and has no share in the fits. In several persons attacked with epilepsy, I have ascertained that the excitations most capable of producing reflex movements did not act more powerfully than in healthy persons, although the experiments were made a short time before a seizure, that is, at a time when the reflex excitability ought to have been at its highest degree. In a young girl, particularly, we have ascertained that tickling the sole of the foot, the axilla, the lips, &c., produced less reflex movements than usual, although she was then expecting a fit, which came on, in fact, about ten minutes afterward. The researches made by Romberg and by Professor Hasse (see his admirable work: *Krankheiten des Nervenapparates*, in Virchow's *Handbuch der Pathologie*, Vol. IV., Part 1st 1855, p. 254) on the production of reflex movements during fits of epilepsy, cannot prove much against or in favor of the existence of a great reflex excitability, or reflex force in epileptics, because if the experiment be made in the beginning of the fit, it is almost impossible to know whether the convulsions result from the experimental excitations, or are normal parts of the fit; and if the experiment is made at the end of the fit, the absence then of reflex movements proves only that the fit has exhausted the vital properties of the muscular and nervous tissues. Hasse concludes, from his own and from Romberg's experiments, that the greatest variety in the energy of reflex phenomena exists during the fits of epilepsy.

Whilst we admit that in epilepsy there is almost always, and perhaps always, an increased reflex excitability, alone or together with an increased reflex force, we admit also that there is, in a great many cases of fits of epilepsy, a special kind of excitation, acting on the nervous centres. There are, therefore, three distinct elements for the production of a fit.

1st. Increase of the *force* of the reflex property;

2d. Increase of the *excitability* of this property;

3d. An excitation of a special nature, or a very violent one.

Of these three elements, the last two are the most frequent, and perhaps, as we have said, the first of these two is essential. As regards the share of a special excitation in the causation of epilepsy, the cases we have related of the cure of this disease by the section of a nerve, by ligatures, &c., show how considerable it may be. But in my animals, we have, in this respect, a better illustration. When the nerves going to the parts of the face and neck, by the irritation of which we are able to cause fits, are laid bare, we find that their irritation does not produce convulsions. If, in these animals, the fits depended only upon an increased reflex excitability of the parts of the nervous centres whence the nerves originate, we should see convulsions follow when we irritate the trunks of these nerves. As there are none, we must admit that when an irritation (and a slight one is often sufficient) to the cutaneous ramifications of these nerves in the skin causes a fit, there is something special in the nature of the excitation springing from these cutaneous nerves. However, there is in my epileptic animals, an increased degree of reflex excitability in the cerebro-spinal axis, as we find, even after the section of the nerves of the face and neck, that they have convulsions sooner, and lasting longer, than in a healthy animal, when we prevent them from breathing for two or three minutes.

A slight increase of the reflex excitability is not usually sufficient alone to cause fits, and such an increase, without epilepsy, often co-exists with great weakness, as is the case in old people, in convalescents, and in persons who have lost a great deal of blood. In all these cases, reflex movements take place easily under the influence of emotions, fright, or even a sudden noise. Many excitable, though healthy men and women have reflex spasms in the act of coition—hence the name given to this act by Sennert, *epilepsia brevis*.

It is very probable that the reflex excitability, or the reflex force, of the nervous centres, or both, are extremely considerable in those persons who have fits of epilepsy for the first time, caused by a slight blow, or some ordinary moral excitement.

We shall not examine what are the parts of the cerebro-spinal axis in which there is an increase of reflex excitability, because what we have said above of the seat of epilepsy shows what are these parts, their seat being nothing but that of the increased reflex excitability, or, in other words, epilepsy consisting chiefly in that increased excitability. If it were proved that epilepsy sometimes exists only because the force of the reflex property is increased, its excitability being normal, we should have to admit that the seat of epilepsy is in almost the whole length of the cerebro-spinal axis, because, as we intend showing elsewhere, the force of the reflex property increases or decreases everywhere at the same time.

We must say, that although we admit that fits of epilepsy depend ordinarily on an increased reflex excitability, frequently combined with the existence of some special kind of irritation originating in the skin, in the mucous membranes, &c., we admit as *possible*, that without any increase of excitability, certain irritations on some parts of the encephalon may produce fits of epilepsy. We well know that the least puncture with a needle or pin of the *processus cerebelli ad pontem*, and, as I have found, of the auditory nerve, and of certain parts of the medulla oblongata, in mammals, is sufficient to produce fits of a peculiar kind of epilepsy, in which the animal rotates around the longitudinal axis of its body, in consequence of the convulsions. In man this kind of epilepsy has been frequently observed, and as the phenomena are the same as in animals (except as regards the duration of the fit, which in man is short, while in animals it lasts *usually* as long as life), it may be that the rotary convulsions have been produced, although there was no increased reflex excitability, in man as it is in animals.

Many discussions have taken place among physicians concerning the first phenomenon of a fit of epilepsy. We are, however, yet to know which of the epileptic phenomena is most frequently the first. Is it the paleness of the face, as Prof. Trousseau and others believe? Is it a spasm of the larynx, as was admitted by Dr. Marshall Hall? Or is it the loss of consciousness? We think there is no doubt that either of these phenomena may be the first, but we do not know which is most commonly the first. They usually take place at the same time, and in some cases they may be entirely missing, or exist only after other phenomena.

Among the most interesting of these ordinarily first phenomena, is the paleness of the face. Delasiauve (*Loco cit.*, pp. 56, 60, 66 and 77), considers it as extremely frequent in all kinds of attacks, from the simple slight absence of mind to the most complete epileptic seizure. Trousseau and Bland Radcliffe (*London Medical Times and Gazette*, March, 1856, p. 303–304), are inclined to consider it as a constant symptom, and also as the first one. This paleness has not been explained. We consider it as a most interesting symptom, as it leads to a very probable explanation of the loss of consciousness in epilepsy. After Prof. Claude Bernard had discovered that the section of the cervical sympathetic nerve is followed by a dilatation of the bloodvessels of the face, I found that when this nerve is irritated by galvanism there is a contraction of these bloodvessels, and I explained the facts discovered by the eminent French physiologist and by myself, by considering the sympathetic as the motor nerve of the bloodvessels of the face. I found, also, that the branches of the sympathetic nerve which animate the bloodvessels of the face, originate from the spinal cord with the branches of the same nerve going to the iris. (See my *Exper. Researches in Physiol. and Pathol.*, 1853, p. 9–10, and p. 75; and the *Medical Examiner*, Aug., 1852,

p. 489.) The theory I then proposed has been almost universally admitted. We have in this theory an easy means of explanation of the paleness of the face in epilepsy. When the excitation takes place in the spinal cord and the basis of the encephalon, which gives rise to the fit, the nerve-fibres which go to the head are irritated, and produce a contraction of its bloodvessels. Of course this contraction expels the blood, and, in consequence, the face becomes pale. Very often another effect, depending on the nerve-fibres of the cervical sympathetic, is produced—the dilatation of the pupil. But the reverse sometimes takes place—a contraction of the pupil occurring, instead of a dilatation. This last phenomenon is easily explained by admitting that the excitation in the nervous centres takes place near the origin of the third and fifth pairs of nerves, and not of that of the cervical sympathetic, as is the case when the pupil dilates. The paleness of the face, and the dilatation of the pupil (when it exists), soon disappear, chiefly in consequence of the obstacle to the venous circulation in the head, and of the state of asphyxia. The cause of the obstacle to the return of blood from the head is not only the contraction of the muscles of the neck, as Dr. Marshall Hall seems to think, but also in the state of the chest. Dr. Russell Reynolds (*Diagnosis of Diseases of the Brain, &c.*, 1855, p. 176) says that he has observed many cases in which the muscles of the neck were quite flaccid, notwithstanding the darkness of the face, and the leaden hue of the body generally.

Among one of the first symptoms of the fit, and as a cause of the cry, there is a spasm of the laryngeal muscles, and a contraction of the expiratory muscles. This contracted state of the chest acts on the heart so as to diminish the force of its beatings, as is the case in the experiment of compressing the chest, made by E. Weber and others, and it acts on the veins, in preventing the circulation in them. Although compressed, and unable to beat freely, the heart quickly recovers an apparently great strength; the blood, losing its oxygen and becoming black, acts as a powerful irritant upon the central organ of circulation, so that palpitations, sometimes very violent, occur. · Nevertheless, the pulse often remains weak, because the quantity of blood sent to the arteries by the heart is smaller than usual, partly on account of the obstacle to the venous circulation.

We think that at nearly the same time, when the origin of the branches of the sympathetic nerve going to the bloodvessels of the face receive an irritation in the beginning of a fit of epilepsy, the origin of the branches of the same and of other nerves, going to the bloodvessels of the brain proper, also receive an irritation. A contraction then occurs in these bloodvessels, and particularly in the small arteries. This contraction expelling the blood, the brain proper loses at once its functions, just as it does in a complete syncope. Now, as it has been well proved by the researches of Kellie, of Abercrombie, of John Reid, of. Henle and of Foltz, that the quantity of liquid in the cranio-spinal cavity cannot change suddenly, it results, that if there is less blood in the brain proper there must be more in the basis of the encephalon and in the spinal cord. In consequence of the impediment to respiration, the blood sent to the encephalon, as well as to other parts of the body, contains but little oxygen, and is charged with carbonic acid, so that the large quantity of blood accumulated in the basis of the encephalon (the medulla oblongata, the pons Varolii, the tubercula quadrigemina, &c.), and in the spinal cord, is endowed in a high degree with the power which I have shown that such blood possesses, i. e., to excite convulsions. It may be, as Henle has supposed, that the basis of the encephalon is also excited to cause convulsions in consequence of the pressure exerted upon it by the accumulation of blood. The spinal cord, also, in all its length, is then excited to produce convulsions by the blood which circulates in it. The grounds on which I base these views are the following.

1st. There is, in the beginning of a complete fit of epilepsy, an irritation of the parts of the nervous centres from which originate the nerve-fibres of the bloodvessels of the brain, and therefore there ought to be a contraction of these vessels. The cervical sympathetic nerve contains not only the nerve-fibres which cause a dilatation of the pupil, and those which produce the contraction of the bloodvessels of the face in the beginning of a fit, but also the nerve-fibres of the bloodvessels of the brain. Prof. Claude Bernard (*Mémoires de la Soc. de Biologie*, for 1853, p. 94) has found that when the cervical sympathetic nerve is divided on one side, the temperature of the brain is increased in the corresponding side. We have shown that this elevation of temperature depends upon the circulation of a larger amount of blood, which is the consequence of the paralysis of the bloodvessels, due to the section of their nerve-fibres. Some experiments of Donders, and of his pupil, Van der Beke Callenfells, have also shown the influence of the sympathetic on the arteries of the pia mater (see Donders' *Physiologie des Menschen*, Leipzig, 1856, p. 138 and 140); they have seen these arteries contract when the sympathetic was irritated.

2d. We have said that we consider as reflex the convulsions of epilepsy, whether they depend on centric or eccentric excitations. The contractions of the bloodvessels of the brain and face in a fit of epilepsy are also reflex. We have proved elsewhere (see my *Exper. Researches in Physiol. and Pathol.*, 1853, p. 34) that bloodvessels may contract by a reflex action, as well as muscles. In experiments with our distinguished friend Dr. Tholozan, Professor at the Military Medical School of Paris, we have found that the bloodvessels of one hand contract, by a reflex action, when the sensitive nerves of the other hand were irritated by being exposed to the influence of water at the freezing point. Schiff (*Comptes Rendus de l'Acad. des Sciences*, vol. xxxix., 1854, p. 509), Donders (*loc cit.*, p. 139), myself, and more recently M. Vulpian (*Gaz. Méd. de Paris*, 1857, p. 18), have found that the bloodvessels of the ear in rabbits contract by reflex action, when the central part of the divided auricular nerve is irritated. I have found, besides, that the splanchnic nerves and other branches of the sympathetic have a reflex action on the bloodvessels of the heart. (See my paper, *Recherches Expérim. sur la Physiol. et la Pathol. des Capsules surrénales*, 1856, p. 30.) All these facts establish beyond doubt that the bloodvessels, as well as the muscles of animal life, may contract by a reflex action. In a seizure of epilepsy, therefore, the bloodvessels of the brain proper, those of the face, the muscles of the neck, of the larynx, &c., may contract by a reflex action, either separately or at the same time.

3d. To say that an explanation is a good one because we do not know or because we cannot imagine any other one, is an argument which rarely has any value; but when the explanation is not only possible, but is even rendered very probable, as is the case with our theory of the loss of consciousness in epilepsy, it is an argument of a positive value that no other theory (except one or two having many facts against them), has been proposed heretofore.

4th. It might be objected to the explanation we propose, that the loss of consciousness is too rapid to be due to a contraction of bloodvessels. There is a fact which answers peremptorily this objection; it is, that when the cervical sympathetic is irritated by a powerful electro-magnetic current, the contraction of the blood-vessels of the face, and particularly of those of the ear, is almost immediate, and so considerable that many of the small arteries seem to expel completely their contents. Now, as everybody knows that even a diminution in the supply of blood to the head, as in ordinary syncope, is sufficient to produce an immediate loss of consciousness, *a fortiori* is it so if the nerve-fibres, irritated in the nervous centres, produce a contraction in the bloodvessels of the brain proper.

5th. As we see that the bloodvessels of the face, after a con-traction of very short duration, dilate and become turgid, it might be asked if it be not so with the bloodvessels of the brain proper, and why, in that case, there is no return of consciousness when the blood returns in the dilating bloodvessels. We answer that it is probable that the cerebral bloodvessels dilate, like those of the face ; but that when this dilatation takes place, the blood which then reaches the brain does not contain oxygen enough, and is charged with too much carbonic acid, to be able to regenerate the lost function of this organ. It is only when the respiration has become almost completely free, that the functions of the brain re-appear.

6th. It might be objected, also, that the theory does not explain why the nerve-fibres going to the bloodvessels of the brain pro-per are excited, while those of the bloodvessels of the base of the encephalon are not. The theory has not to explain this differ-ence ; *it is a fact* that the action of the brain proper is lost, while the action of the basis of the encephalon is very much increased, during a fit of epilepsy ; and all that the theory has to do is to explain the loss of action in one part, and the cause of increased action in another. However, we may add that if the bloodvessels of the base of the encephalon are not excited to contract, it is, ac-cording to all probability, because their nerves originate in another place from those of the cerebral bloodvessels ; and as we know that the nerves going to certain muscles are excited in the begin-ning of a fit, while others are not, we may understand easily that the same thing exists for the nerves of the various encephalic bloodvessels.

7th. As regards the influence of blood charged with carbonic acid on the nervous centres, we will refer to our often-quoted work (p. 80, and p. 101–124) ; and we will merely say here that we have found that the injection of blood charged with carbonic

9

acid into the carotid or into the vertebral arteries, at once causes epileptiform convulsions.

8th. It might be objected that the bloodvessels of the base of the encephalon and of the spinal cord ought to be excited to contract by two causes after the fit has lasted some time; the first cause being the excitation of all the parts of the cerebro-spinal axis in which there is blood charged with carbonic acid, and, consequently, the excitation of the nerves of the bloodvessels of the basis of the encephalon, because these nerves take their origin somewhere in those excited parts of the cerebro-spinal axis; the second cause being the direct excitation of the smooth muscular fibres of the bloodvessels of the encephalon by the blood charged with carbonic acid. Now if the bloodvessels contract, whether it is on account of the first or of the second cause, or of both, it seems that the fit ought to be diminished at once. But in the first place, it is probable that the bloodvessels contract irregularly, some at one time, some at another. In the second place, blood charged with carbonic acid, after its first action (which is an excitation) has a secondary action, which causes the loss of the contractility of the muscular layer of the bloodvessels. In the third place, the obstacle to the return of venous blood may cause the bloodvessels to dilate to such an extent that they cannot contract, as is the case with the heart when its cavities are too full.

9th. If there are contractions in the bloodvessels of the **brain** proper, as there are in the muscles of animal life, in the beginning of an epileptic seizure, it is very easy to explain the variety of sensorial and other cerebral symptoms of epilepsy. In the same way as there are *certain* muscles that contract in the neck, in the larynx, or elsewhere, we may admit that there are *certain* bloodvessels that contract either in some parts of the brain proper, or in the nervous portions of the organs of sense, and in consequence, there is a trouble or loss of either one or several senses, or of the intellectual faculties, consciousness remaining more or less entire; or there is a successive loss of sight, of hearing, of the intellectual faculties, and, at last, of consciousness.

10th. It is well known that sometimes the compression of the carotid arteries stops a fit of epilepsy. Cases of this kind have been mentioned by Liston, Earle, Albers, &c. The same operation in certain animals, and particularly in rabbits in good health, is sometimes sufficient to cause convulsions, so that we are led to the question, How can the same circumstance in one case cause convulsions, and in another diminish or destroy them? My theory may give an explanation of this apparent opposition. Changes in the quantity of fluid in the cranio-spinal cavity cannot take place suddenly, and if there is a considerable diminution in the quantity of blood which enters this cavity, as is the case when the carotid arteries are compressed, there is necessarily a corresponding diminution in the quantity that goes out. The blood which reaches the encephalon by the vertebral arteries having to fill a much larger space, circulates more slowly and becomes much more charged with carbonic acid, and, besides, furnishes much less oxygen to the encephalon, so that if the compression of the carotid arteries be made in healthy animals, it causes convulsions, just as I have found that blood much charged with this acid injected into the carotid arteries, causes convulsions; whereas, if the compression of these arteries be made in man, during an epileptic seizure, there is at first usually a momentary increase in the intensity of the fit, and sometimes after one or two minutes, rarely sooner, a diminution in the violence of the convulsions, and in some cases, a complete cessation of these contractions. Those who have observed what takes place in animals when they are asphyxiated, have remarked that after violent convulsive struggles, while the blood is becoming more and more charged with carbonic acid, there is a diminution of the convulsions, and at last nothing but rare respiratory efforts. Carbonic acid, after having excited the vital properties of the nervous system, seems to destroy them gradually, allowing for a time, however, the production of respiratory movements. The compression of the carotid arteries in epileptics, during a fit, induces a state of asphyxia greater than that already existing, and in so doing, diminishes the vital properties so much that there are no more convulsions. Respiration taking place* then, and the bloodvessels of the brain proper relaxing, the whole encephalon receives more oxygenated blood, and the patient recovers in the same way, and by the same means, that he does when the compression of the carotid is not employed in a fit.

---

* Of all the reflex phenomena, the regular inspiratory and expiratory movements are those which last the longest; it is so during agony resulting from any disease, it is so after chloroform or ether have been inhaled in large doses, it is so in asphyxia by hanging, drowning, &c., and it is so also in epilepsy.

The theory of epilepsy that we have arrived at from the examination of the phenomena of this disease, is not in opposition with any that we know; and, still more, we might easily show that it is in harmony with the most important facts concerning the causes, the variations of the symptoms, the consequences and the treatment of this convulsive affection. We will merely point out, in addition to what we have related above, that the production of epilepsy by lead (which is an excitant of contraction in bloodvessels), by loss of blood, &c., and the important relations of epilepsy with intermittent fever, are facts in perfect harmony with our theory.

We must now say a few words, 1st, on the production of the change in the cerebro-spinal axis, which chiefly constitutes epilepsy (*i. e.*, the augmentation of the reflex excitability); 2d, on the production of the change of certain parts of the skin, mucous membrane, &c., which renders these parts capable of exciting epileptic seizures; 3d, on the mode of production of a fit of epilepsy from excitations springing either from a peripheric part or a central part of the nervous system; 4th, on the consequences of an epileptic seizure, and on the inter-paroxysmal state.

1st. The production of a change in the reflex excitability of the cerebro-spinal axis we think may take place in two different ways, one of which is a *direct* abnormal nutrition, as in syphilitic, scrofulous or rheumatic epilepsy, while the other is an *indirect* abnormal nutrition, due to some excitation from a peripheric or a central part of the nervous system. The *modus operandi* of such excitations we do not know positively, but very likely, in a number of cases, at least, it is through the bloodvessels of the cerebro-spinal axis that these excitations operate to change the nutrition of this nervous axis. We have ascertained that many substances which act upon the spinal cord, either in increasing its reflex faculty (such are strychnia, morphia, &c.), or in diminishing it (such are belladonna, ergot of rye, &c.), produce their effect chiefly by their influence on the bloodvessels of this nervous centre. When they excite the bloodvessels to contract, they diminish nutrition, and cause paralysis; when they diminish the contractility of the bloodvessels, and therefore allow them to dilate, there is more blood in the spinal cord, and its nutrition is increased. Then the reflex faculty becomes greater, and irritations may cause convulsions. In animals and men, not having taken any of these substances, the reflex excitability of the cerebro-spinal axis may be increased in

the following ways. An excitation on some part of the nervous
system causes a contraction of the small bloodvessels of a part of
the cerebro-spinal axis, and as the same quantity of blood still ar-
rives by the various arteries in the cerebro-spinal cavity, it results
that if the small ramifications of some arterial branches are contracted, the others receive more blood, so that nutrition, and, in
consequence, the reflex excitability, augment in the parts to which
they are distributed. But this is not likely to be the most frequent
mode of increase of nutrition. We have found that when a vascu-
lar nerve is excited for a long while, the contraction of the bloodvessels after a certain time ceases, and a dilatation takes place,
which lasts longer than the contraction, although the nerve is still
excited : this is a paralysis by excess of action. In the nervous
centres, very likely this paralysis of the bloodvessels supervenes
also after considerable contractions, and in consequence of this
paralysis, nutrition is increased in the parts of these centres where
it exists, as we have found that nutrition is increased in the nerves and
muscles of the face, when their bloodvessels are paralyzed. With
the increase of nutrition in the nervous centres comes the augmentation of the reflex excitability, which seems to be the principal
element of epilepsy.

Besides these causes, there is another of greater importance,
which may exist when they do not : the nerve-fibres animating the
bloodvessels of the parts of the cerebro-spinal axis where epilepsy
has its seat, may be paralyzed, as the nerve-fibres of the muscles
of animal life are, by a disease of some part of the nervous centres,
and the consequence of this paralysis is necessarily an increase of
nutrition and of reflex excitability. This is a fact which we have
positively ascertained ; the section of a lateral half of either the
medulla oblongata or the spinal cord is the cause of paralysis of
the bloodvessels of the cord on the same side, the consequence of
which paralysis is that nutrition and the reflex excitability of the
cord become much increased. When the spinal cord is cut across
entirely, in mammals as well as in cold-blooded animals, the part
separated from the encephalon has its bloodvessels paralyzed, and
therefore dilated. Nutrition and the reflex excitability are soon
much increased in this part, and it is sufficient to touch the skin or
the mucous membrane of the genital organs, or of the anus, to de-

termine violent spasms.* This cause of production of epilepsy, or at least of an increased reflex excitability, must exist in a very great degree in cases of tumors of the pons Varolii, or of the medulla oblongata, and if they do not cause this convulsive affection more often it is very probably because the moral and the emotional excitation of fits cannot act in many of these cases.

When an excitation coming from some peripheric nerve produces in the cerebro-spinal axis the change of nutrition which causes epilepsy, it is very likely that this excitation sometimes, if not always, acts otherwise than by producing a contraction of some bloodvessels. Whether this action is like those due to electricity or not, we cannot tell, but we think that an opinion which we had held for many years with Donders, and some other physiologists,† must be modified. This opinion is that all the nervous influences on nutrition, secretion, &c., either direct or by reflex action, act only in causing contractions or paralytic dilatations of bloodvessels. This view, which has been criticised with much ability by Prof. James Paget, in his admirable lectures on nutrition and on inflammation, seems to have been proved to be too absolute by the important researches of Prof. Ludwig and his pupils (see *Physiol. des Menschen*, von Donders, vol. i., p. 187–9, 1856), which appear to establish positively that there is another mode of influence of the nervous system, at least on certain glands; an influence resembling that possessed by electricity in causing chemical combinations or decompositions.‡

2d. The changes produced in peripheric parts, rendering them able to excite fits of epilepsy, consist more in alterations in the

---

* The same thing sometimes occurs in man. In a case of fracture of the spine, recorded by Dr. Knapp (N. Y. Journal of Medicine, Sept., 1851, p. 198), there was paralysis of the abdominal limbs. A month after the accident, there were slight spasms in those limbs; in four months, the spasms became violent; on exposure to the cold air, or to a sudden touch, the muscles were thrown into the most violent agitation.

† Prof. Claude Bernard, in announcing recently his important discovery of the substance which in the liver gives origin to sugar, expresses himself very strongly in favor of this opinion. (*Gaz. Méd. de Paris*, 1857, p. 202.)

‡ We still maintain, however, as we have done for many years, that the influence of the nervous system on nutrition and secretion, either direct or reflex, is in a great measure due to the influence of nerves on the muscular layer of the bloodvessels. Even galvanism, in improving nutrition, we have proved to act partly in this way; it contracts the bloodvessels, and in so doing diminishes circulation and warmth. But after a certain time of violent contraction, the bloodvessels become paralyzed and dilated, so that more blood passes through them, and the temperature and nutrition are increased.

nature of the excitations that may spring from peripheric nerves than from an increase in the felt excitations coming from these nerves. We have shown already that in our animals the skin is not more sensitive in the parts of the face which are capable of exciting fits than in the other parts of the face which have not that power (see § IV.). In man, as we have also shown elsewhere (see § XI.), it is to the nature of the excitation, and not to the degree of the pain, springing from some peripheric nerve, that we must attribute the production of the fits. The fact that excitations, starting from the periphery and causing fits, may not be felt, and the fact that when there are sensations accompanying these unfelt excitations, they may vary as to their kind, and sometimes be very feeble, certainly are important arguments to show that the real exciting cause of the fit is something which is not felt. If the term *aura epileptica* had not been employed already to express the sensations which accompany the excitation of the fits, it would be well to employ it to name the unfelt excitation which is the real exciting cause.

In inquiring into the nature of the unfelt aura, we find that very probably it is nothing but a violent excitation originating in the excito-motory nerve-fibres. Dr. Marshall Hall and Mr. Grainger have long ago imagined that there are nerve-fibres which are employed in reflex actions, and not in sensations and in voluntary movements; but they did not adduce direct facts to prove the correctness of their views. I have found many facts which seem to give the proofs hitherto needed that there are nerve-fibres which are employed in exciting reflex actions, and which are neither sensitive nor capable of transmitting sensitive impressions to the encephalon. I have found also, that the excito-motory power, like the sensibility of nerves, varies in different parts of their length (see my *Experimental Researches applied to Physiology and Pathology*, New York, 1853, p. 98), and also in the same part, according to various circumstances.

Besides, I have ascertained that in certain parts where the excito-motory power seems not to exist, it may be generated, and become considerable. Now, as the fibres which have this power seem not to be sensitive, we understand why an excitation may originate from them, reach the nervous centres, produce the loss of consciousness and convulsions by a reflex action, without giving pain, or even any sensation. We may understand also that this reflex excitation may produce cramps by a reflex action in the muscles which are in the neighborhood of the starting point of the excitation, which cramps give rise to a pain wrongly considered as a primitive aura, although it is only a secondary and almost inefficient one.

With the view that in the very beginning of epileptic fits, caused by excitations coming from peripheric nerves, it is not the sensitive nerve-fibres, but only the excito-motory fibres which are in action, we can easily explain many facts. For instance, in my animals, the power of giving rise to fits, belonging to the cutaneous ramifications of nerves and not to their branches or trunks; in man, the absence of sensations, although there is an excitation from some peripheric nerves, as in the case of M. Pontier (see § IX., Case VII., and many others mentioned in § XI.).

What the causes of the increase of the excito-motory power are, we cannot tell positively. We know, however, that some causes increase all the vital properties of nerves everywhere, and among these causes we will point out a paralysis of the bloodvessels, or the development of inflammation. But there are other causes of which we are ignorant; in my animals, for instance, there is but a slight increase in the vascularization of the part of the skin which has the power of giving rise to fits, and this might be due to the pinching employed to irritate the skin.*

The changes taking place in the peripheric nerves, either in the skin, in the mucous membranes or in their trunks, when they become able to excite epileptic fits, may be produced by the influence of distant parts. For instance, in my animals, alterations of the spinal cord as low down as the *cauda equina* have sometimes been productive of the peculiar change in the face and neck which renders these parts able to excite fits. In man, tumors of the brain seem to have produced a similar change in one arm.

In my animals I cannot decide whether it is through some direct nervous influence upon the nutrition of the skin of the face and neck, or if it is through an indirect influence, and by means of the bloodvessels, that the spinal cord acts on this part. I have found

---

* I had said, in a paper read last year at the *Académie des Sciences* of Paris (*Arch. Gén. de Méd.*, Fév., 1856), that in making the autopsy of my epileptic animals, a congestion of the base of the encephalon and of the Gasserian ganglion is found; but I have ascertained since that in a great measure this congestion is a result of the fits and of the irritation of the skin of the face by pinching or otherwise, and not a circumstance preceding the first fit, and connected with the production of the increase of the excito-motory power of the skin.

that changes in nutrition occur in other parts of the head—such as
the cornea—in animals upon which the section of a lateral half of
the spinal cord has been made, but is this a direct or an indirect
influence? I cannot decide. It is very well known that the sympathetic nerve in the abdomen may influence the nutrition of the
eye through the spinal cord, but does the influence result from a
change in the calibre of the bloodvessels of the eye, or is it a direct influence, like that of certain nerves on the salivary glands,
according to the great discovery of Ludwig?

As regards tumors of the brain, the important case of Odier
(see § VIII., Case I.) seems to show that they may produce in the
arm that peculiar change in peripheric nerves which renders them
able to excite fits of epilepsy. But it is by far much more probable that it was not by an action of the brain, but through the irritation of the sensitive or excito-motory nerves of the scalp, or in
consequence of the compression of the base of the encephalon,
that the change of nutrition took place in the arm.

3d. In the two preceding sections I have examined how are produced the two organic causes of epilepsy; i. e., the increase of the
reflex excitability of certain parts of the cerebro-spinal axis, and
the increase in the excito-motory power of the peripheric nerves.
I have now to say a few words on the mode of production of the
most interesting phenomena of a complete fit of epilepsy.

The first phenomenon of such a fit is not always the same, and
this explains why the best observers do not agree in this respect.
Dr. Marshall Hall for a long while considered as the first symptom
a distortion of the eye-balls and of the features, and he admitted
as the second phenomenon a forcible closure of the larynx, and an
expiratory effort (*Diseases and Derangements of the Nervous
System*, 1841, p. 323). In many subsequent publications (see
Lancet, June 12, 1847, p. 611, and *Aperçu du Système Spinal*,
1855, p. 101) he seems to consider as the first phenomena the contractions of the muscles of the neck and of the larynx. Dr. C.
J. B. Williams (*General Pathology*, 2d Am. Ed., p. 166) says
that the first phenomenon is a palpitation of the heart. Herpin
(*Loco cit.*, p. 421-5) after having tried to show that when there is
an aura the first phenomenon consists in a local cramp, says that
the second phenomenon (the first when there is no aura) is the epileptic cry. According to Beau (*Arch. Gén. de Méd.*, 1836, p.
339), Delasiauve (*Loco cit.*, p. 65) and Hasse (*Krankheiten des

Nervenapparates, 1855, p. 251), the epileptic cry, in the most complete cases of epilepsy, may not exist. I have witnessed two fits
of epilepsy·in which the most violent convulsions and a complete
loss of consciousness, followed by coma, took place without cries.
Is the loss of consciousness the first symptom? Most of the principal writers, who ignore the power of the reflex actions, consider
the cry as a proof of feeling: surprise, according to Beau; surprise and pain, according to Herpin *(Loco cit.,* p. 477); surprise,
convulsion and pain, according to Delasiauve *(Loco cit.,* p. 77), and
they admit, therefore, that the loss of consciousness is not the
first symptom, at least in most cases. Billod attributes the cry to
the convulsive spasm of the laryngeal muscles, and to a convulsive expiration *(Annales Méd. Psychol.,* Nov. 1843). According
to him, the loss of consciousness precedes the cry, which is not a
symptom of surprise or of pain. Hasse considers the cry as being
probably the result of a reflex action *(Loco cit.,* p. 251–2). I have
tried to show elsewhere *(Exper. Researches applied to Physiol.
and Pathol.,* New York, 1853, p. 54–5) that cries in animals or in
children deprived of their brain, may be due to a mere reflex' action; the vocal cords being contracted, and the expiratory muscles expelling quickly the air contained in the chest, the sound
which we call a cry is produced. In epilepsy, the loss of consciousness, which is equivalent to the loss of the brain, allows a cry
to take place by reflex action. In the most complete and violent fits
of epilepsy, we think that the first phenomena are almost always,
1st, the contraction of the bloodvessels of the face, which causes
the paleness, noted particularly by Prof. Trousseau, by Delasiauve
and by Dr. Bland Radcliffe; 2d, the contraction of the bloodvessels of the brain proper, which causes the loss of consciousness.
The cry, either at the same time, or immediately after, is produced
by the spasmodic contraction of the expiratory muscles, driving
the air forcibly through a contracted glottis. At the same time,
also, almost always some muscles of the face, of the eye and of
the neck contract. Sometimes, also, the spasm extends at once to
the muscles of the upper limbs, and afterward to the whole body.
All these phenomena are sometimes produced at once, and all are
the results of an excitation springing from some part of the excito-motory side of the nervous system. In other cases there is an
evident succession in these phenomena; the paleness of the face
and the loss of consciousness (both resulting from contractions of

the bloodvessels) take place at first, with some spasmodic actions of the muscles of the eye and face, and then come the cry and the tonic contraction of the muscles of the limbs and trunk..

The following table will show how the principal phenomena are generated, *one by the other*, in the most common form of the violent and complete epileptic seizures.

CAUSES.	EFFECTS.
1. Excitation of certain parts of the excito-motory side of the nervous system.	1. Contraction of bloodvessels of the brain proper and of the face, and tonic spasm of some muscles of the eye and face.
2. Contraction of the bloodvessels of the face.	2. Paleness of the face.
3. Contraction of the bloodvessels of the brain proper.	3. Loss of consciousness, and accumulation of blood in the base of the encephalon and in the spinal cord.
4. Extension of the excitation of the excito-motory side of the nervous system.	4. Tonic contraction of the laryngeal, the cervical and the expiratory muscles (laryngismus and trachelismus).
5. Tonic contraction of the laryngeal and of the expiratory muscles.	5. Cry.
6. Farther extension of the excitation of the excito-motory side of the nervous system.	6. Tonic contractions, extending to most of the muscles of the trunk and limbs.
7. Loss of consciousness, and tonic contraction of the trunk and limbs.	7. Fall.
8. Laryngismus, trachelismus, and the fixed state of expiration of the chest.	8. Insufficient oxygenation of the blood, and general obstacle to the entrance of venous blood in the chest, and special obstacle to its return from the head and spinal canal.
9. Insufficient oxygenation of the blood, and many causes of rapid consumption of the little oxygen absorbed, and detention of venous blood in the nervous centres.	9. Asphyxia.
10. Asphyxia, and perhaps a mechanical excitation of the base of the encephalon.	10 *Clonic convulsions everywhere*, contractions of the bowels; of the bladder; of the uterus; erection; ejaculation; increase of many secretions; efforts at inspiration.
11. Exhaustion of nervous power generally, and of reflex excitability particularly, except for respiration. Return of regular inspirations and expirations.	11. Cessation of the fit; coma or fatigue; headache; sleep.

We have but little to say in explanation of the above table, which only gives, as we hardly need to remark, a type of a complete seizure.

Writers on epilepsy are unanimous in considering the *fall* as due only to convulsions, while it is certainly, in a measure, the consequence of the loss of consciousness, which alone causes it in some cases of epileptic vertigo without convulsions.

We do not think that laryngismus in epilepsy has the immense importance given to it by Dr. Marshall Hall. In the first place, in persons in whom the reflex excitability is not increased, laryngismus exists frequently, in whooping cough, in asthma, &c., without producing epileptic convulsions. In the second place, epileptic convulsions may exist before laryngismus (Hasse, *loco cit.*, p. 252). If, instead of saying that laryngismus is the essential cause of convulsions in a fit of epilepsy, we say that asphyxia, whether produced by laryngismus or by other causes, is the source of a certain part of the convulsions in the violent and complete fits of epilepsy, we shall be much nearer the truth. If we say also that laryngismus is nothing but a spasm of certain muscles—spasm produced by a reflex action at the same time that there are other spasms in the bloodvessels of the brain proper, of the face, and also sometimes of the whole surface of the body, and in the muscles of the head, of the trunk and limbs, and that all these spasms are reflex contractions, due to the same excitation, we shall be much nearer the truth than by admitting Dr. Hall's views.

Not only is it wrong to say that convulsions in epilepsy are due only to laryngismus, but it would be wrong also to say that they are due only to asphyxia, whatever be its cause. The tonic convulsions, which, according to Dr. Copeland (*Dict. of Med.*, vol. i., p. 786), and to Herpin (*Loco cit.*, p. 451,) always exist in the beginning of fits of epilepsy, are not to be attributed to asphyxia, neither are the convulsive rotary movements which sometimes exist, and which result principally from the irritation of some parts of the isthmus of the encephalon. The tonic convulsions may occur in almost all the muscles of the body at once, simulating tetanus, or they come first in the larynx, the neck, the eyes, or the face, and thence extend to the upper limbs, and at last to the trunk and inferior limbs. These convulsions are mere reflex spasms, as are the contractions of the bloodvessels. Their duration is only of some seconds, according to Copeland (*Loco cit.*, p. 786), or a quarter of a minute, according to Herpin; but they may appear again during the seizure, as Hasse (p. 252) and Herpin (p. 430) justly observe, and as I have twice seen. This kind of convulsion, and also the rotary convulsions, cannot be the result of laryngismus, because asphyxia does not seem able to produce them. Asphyxia causes only clonic convulsions, and it seems that we must attribute to it the universal clonic convulsions of a complete fit of epilepsy. We have perused the history of many hundred cases of epilepsy,

and we have witnessed eight violent fits in as many epileptics; and
in all these cases, universal clonic convulsions have begun only after the appearance of symptoms of asphyxia. In healthy animals
prevented from breathing, clonic convulsions begin in less than half
a minute, and they are universal and very violent in about three
quarters of a minute. General clonic convulsions are produced
sooner, i. e., in twenty to thirty seconds, rarely later in my epileptic animals, when they are absolutely deprived of respiration.

If universal clonic convulsions in epilepsy seem to be due only
to asphyxia, the same thing cannot be said of local clonic convulsions, which frequently occur before there is a sufficient degree of
asphyxia to produce them. For more than six years I have, almost
every day, seen in my animals local clonic convulsions following a
tonic spasm of the muscles of the face and neck, long before the
time when a complete deprivation of breathing, had it existed,
could have produced convulsions.

We may conclude—1st, that neither the general nor local tonic
spasms, nor the local clonic convulsions of epilepsy depend upon
asphyxia, and that therefore they are independent of laryngismus;
2d, that asphyxia in epilepsy does not usually depend upon laryngismus alone, but that it may result from many other causes, such
as the spasmodic contraction of the muscles of expiration, or from
alternate contractions and relaxations of all the muscles of the chest
and diaphragm. Asphyxia is also partly due to the accumulation of
black blood in the nervous centres, by the obstacle to the return
of venous blood; and partly also because the energetic actions of
the nervous centres and of the muscles cause a rapid consumption
of oxygen, and charge the blood with carbonic acid. The experiments of Roupell (British Assoc., 1841; see *Am. Jour. of Med.
Sciences*, January, 1842, p. 243) show conclusively the influence of
carbonic acid in causing clonic convulsions, with foam at the mouth,
&c., as in epilepsy.

Asphyxia is not only the cause of the most violent general clonic
convulsions in a fit of epilepsy, but it is also the usual cause of the
contractions of the bladder, of the bowels, of the uterus, and of the
muscles which produce the erection of the penis and the ejaculation. We admit, however, that all these muscular contractions
may be produced by the same cause to which are due the first tonic
spasms, and that they exist sometimes when there is no considerable asphyxia, or before the beginning of asphyxia.

We do not pretend to give here an account of all the phenomena of epilepsy: we abstain from speaking of those which have been already well explained, such as relate to the tongue, and also to the coma so frequent after a very violent fit. We do not need to say that an immense variety of these phenomena may be observed in epileptics, and that this variety depends upon the part first excited in the nervous centres, and upon the degree of the reflex excitability and of the reflex force of these centres.

In the beginning of a fit of epilepsy, it sometimes happens that the heart's action is stopped more or less completely. This stoppage may be due to two essentially different causes: 1st, the heart being compressed by the spasmodic contraction of the chest, is mechanically rendered more or less completely unable to move, as may be the case even in a healthy man, according to the interesting experiments made by Ed. Weber (see *Müller's Archiv.*, 1851, p. 88) upon himself and upon other persons; 2d, a reflex action upon the bloodvessels of the heart may determine their contraction, and therefore stop at once the movements of this organ, in the same way that they are stopped sometimes by an emotion, by chloroform, by an irritation of the abdominal sympathetic or other nerves, &c. It is well known that the quantity of many secretions, or their quality, may be altered during a fit of epilepsy. These changes may be due to at least two distinct causes: 1st, there may be a reflex influence upon the various glands, or upon their bloodvessels,* as there is a reflex action upon the bloodvessels of the face, and very probably of the brain; 2d, asphyxia is certainly one of the causes of the changes in secretions during an epileptic seizure. (See my *Experimental Researches applied to Physiol. and Pathol.*, 1853, p. 113–114.) As regards *saliva*, Lehmann mentions that there is a great flow of it in horses which

---

* A discussion of priority took place, some time ago, between Dr. H. F. Campbell, of Augusta, Ga., and Dr. Marshall Hall, concerning the discovery of the existence of *reflex* secretions. There is no doubt that Dr. Campbell, who published his first paper in May, 1850, has the priority over Dr. Hall, but these two able physicians seem not to know that many of the German writers had long ago gone very far in that field of the reflex secretions. As regards the normal reflex secretions, I will point out a short note which I published in 1849 (*Comptes Rendus de la Soc. de Biologie*, p. 104, July, 1849, and *Gazette Méd. de Paris*, 1849, p. 644), in which, besides the citation of many facts, I mentioned particularly the production of sweat by a reflex influence. As regards the pathological reflex secretions, I will direct the reader particularly to the various works of Henle, published in 1840, in 1841, and later. I will add, that the question is not now whether there are such things as reflex secretions, but whether, in the reflex secretions and in the reflex changes of nutrition, there is only an influence upon the muscular elements of the bloodvessels, or if there is some special electric or nervous influence. (See the treatises by Prof. Ludwig, by Prof. Donders, and by O. Funke; and the great work of Spiess, *Pathologishe Physiologie*, 1857.)

breathe for a few minutes atmospheric air containing ten per cent.
of carbonic acid (*Physiol. Chemistry*, English translation, 1853,
vol. ii., p. 177). In the experiments of Roupell, already quoted,
there was much foam at the mouth in dogs, after injections of carbonic acid into their veins.

We must say a few words more to explain the relations of epilepsy with sleep, and with the loss of blood. It is well known
that sleep is a very favorable condition for epileptic seizures; indeed, we may say that in many persons who are not, however, epileptic, sleep is a slight attack of epilepsy. The loss of consciousness, of course, exists, and convulsions in many muscles are very
frequent. Whatever be the nature of sleep, it is quite certain that
it is a state of sub-asphyxia, and in this respect also it resembles
epilepsy. It is certain, also, that the circulation in the encephalon is modified in both epilepsy and sleep. But the kind of trouble in the encephalic circulation, in the beginning of an epileptic
seizure, is not the same as that which takes place during sleep.
In epilepsy, according to the theory we have proposed, there is at
first a contraction of the bloodvessels of the brain proper, and it
is only after a fit has existed a few seconds, or a little longer, that
the spasm of these vessels ceasing, circulation of rather black
blood takes place in them, as during sleep. We have said above
that if the carotid arteries are compressed during a seizure of epilepsy, and if the fit is stopped by it, the reason of this influence is
to be found in an increase of the pre-existing asphyxia. We find
an interesting fact in harmony with this view, in a short paper by
Prof. A. Fleming, of Cork, who states that sleep is easily produced in persons who are not epileptic, by the compression of the carotid arteries. (*British and Foreign Med.-Chir. Review*, April,
1855, p. 404, Am. ed.) The diminution in the supply of arterial
blood produces the same effect as the obstacle to the return of
venous blood, which obstacle is known to cause sleep. When the
carotid arteries do not send blood to the brain proper, circulation
is almost stopped in that part of the encephalon, and the absence
of oxygen produces a sudden paralysis of the brain, and after a
few seconds there is a state of slight asphyxia, marked by sterto-
rous breathing. (Fleming.) If attacks of epilepsy take place
more easily during sleep than during wakefulness, it seems that it
is on account of the slight state of asphyxia existing during sleep.

In cases of epilepsy, or, at least, of convulsions, due to a loss of blood, or to insufficiency in the quantity of this liquid, there is also, as a cause of the fits, a state of asphyxia, due to the fact that as there is less blood reaching the cranio-spinal cavity, the circulation is slower in the nervous centres, and the blood has time to become charged with carbonic acid and to become an excitant. Besides, it is certain that when there is not blood enough circulating in the nervous centres, their reflex excitability becomes increased at the same time that their *reflex force* diminishes. The asphyxia due to a diminution of blood seems to cause both the state of the nervous system favorable to the production of epileptic fits, and the excitation which determines them. In the same way, the asphyxia due to various causes during a fit of epilepsy prepares new fits for the future, and actually causes clonic convulsions.

4th. As regards the last question we have to examine, which relates to the *effects* of attacks of epilepsy, we will only say that they depend upon three circumstances : 1st, the absence or great diminution of circulation in the brain proper in the beginning of a fit of epilepsy ; 2d, the circulation of black blood through the nervous centres ; 3d, the pressure upon many parts of the base of the encephalon and of the spinal cord, by the accumulation of blood in their vessels.

In consequence of these causes, various disorders of the mind, of the senses, and of the vital properties of the nervous centres, are produced. We will not speak of these disorders here, and will merely refer the reader to the analysis given of most of them by Dr. Russell Reynolds, in his important researches on the inter-paroxysmal state of epilepsy. (See *Diagnosis of Diseases of the Brain, &c.*, 1855, p. 175, and the *London Lancet*, 1856.)

§ XV. *Treatment of Epilepsy.*—Proposing to develop fully this subject elsewhere, we will merely lay down here a few propositions.

1. The first thing to be done in a case of epilepsy is to find out if its origin is peripheric. The state of all the organs must be inquired into as completely as possible. For some of the means to be employed to detect the peripheric origin of fits of epilepsy, we will refer to § XI.

2. If it be ascertained that epilepsy is of peripheric origin, employ proper means to separate the nervous centres from this ori-

gin, or to remove the cause of the excitation entirely. Leaving aside what relates to the viscera, the application of ligatures, as we have shown in § IX., ought to be tried first. Sometimes it happens, as in a very curious case recorded by Récamier, that the aura will disappear from a place, and re-appear in another; it will be well to pursue it thither, and apply ligatures in the new place.

3. If ligatures fail, this is no reason for despairing of other means having the same object. The nerve animating either the part of the skin from which originates the aura, or the muscle or muscles which are the first convulsed, must be laid bare, and sulphuric ether thrown upon it. This might, perhaps, be sufficient to cure the affection; if it is not, then the nerve must be divided.*

4. The amputation of a limb for epilepsy is a barbarous act, the section of the nerves being all that is necessary.

5. Sometimes blisters, setons, caustics, &c., in the neighborhood of a part which is the origin of an aura, may be sufficient to cure, but these means have not the same efficacy as the application of a red-hot iron.

6. The best means of treating epilepsy seem to consist in the application of a series of moxas along the spine, and particularly the nape of the neck.

7. The nutrition of the nervous centres may be modified, and thereby epilepsy be cured, principally by the medicines which act on the bloodvessels, such as strychnia, but particularly by those which determine contractions in these vessels, such as atropia, ergot of rye, &c.

8. Trepanning, in cases where a blow on the head or some other circumstance seems to indicate it, ought not to be resorted to until cauterization and other means of producing a modification of the conditions of the skin of the head have failed. (See § IX.)

9. Cauterization of the mucous membrane of the larynx, which has been successful in some cases in which there was considerable laryngismus, is an excellent means, not only of diminishing or preventing the spasm of the larynx, but as a mode of producing a modification in the nutrition of the medulla oblongata.

10. As a means of treatment too much neglected, we will point out the possibility of the transformation of epilepsy into intermit-

* We proposed, many years ago, to employ ether instead of the section of the nerves, in traumatic tetanus; this simple treatment will prove more useful for tetanus than for epilepsy.

12